Olaf Kaltmeier

National Parks from North to South

An Entangled History of Conservation and Colonization in Argentina

INTER-AMERICAN STUDIES
Cultures – Societies – History

ESTUDIOS INTERAMERICANOS
Culturas – Sociedades – Historia

Volume 34

Olaf Kaltmeier

National Parks from North to South

An Entangled History of Conservation and Colonization in Argentina

Wissenschaftlicher Verlag Trier

Copublished by

UNO University of New Orleans Press

National Parks from North to South.
An Entangled History of Conservation and Colonization in Argentina /
Olaf Kaltmeier. –
(Inter-American Studies | Estudios Interamericanos; 34)
Trier: WVT Wissenschaftlicher Verlag Trier, 2021
ISBN 978-3-86821-834-3
New Orleans, LA: University of New Orleans Press, 2021
ISBN 978-1-60801-204-6

SPONSORED BY THE
Federal Ministry of Education and Research

The project, on which this book is based, has been funded by the German Federal Ministry of Education and Research (Bundesministerium für Bildung und Forschung, BMBF). The responsibility for the content of this publication lies with the author.

Cover Image: Picture detail from FF.CC. 1929. *Lagos andinos: Parque Nacional del Sud: Montañas – bosques – cascadas*. Buenos Aires.

Cover Design: Brigitta Disseldorf

Library of Congress Cataloging-in-Publication Data
Names: Kaltmeier, Olaf, 1970- author.
Title: National parks from north to south : an entangled history of conservation and colonization in Argentina / Olaf Kaltmeier.
Description: New Orleans, LA : University of New Orleans Press ; Trier, Germany : Wissenschaftlicher Verlag Trier, 2021. | Series: Inter-American studies : cultures - societies - history = Estudios interamericanos : cultures - sociedades - historia ; Volume 34 | Includes bibliographical references.
Identifiers: LCCN 2020044536 | ISBN 9781608012046 (paperback)
Subjects: LCSH: National parks and reserves--Argentina--History. | Nature conservation--Argentina--History.
Classification: LCC SB484.A7 K35 2021 | DDC 363.6/80982--dc23
LC record available at https://lccn.loc.gov/2020044536

Publisher: WVT Wissenschaftlicher Verlag Trier, Postfach 4005, D-54230 Trier, Bergstraße 27, D-54295 Trier, Tel. 0049 651 41503, Fax 41504, www.wvttrier.de, wvt@wvttrier.de

Copublisher: University of New Orleans Press, 2000 Lakeshore Drive, Earl K. Long Library, Room 221, New Orleans, LA 70148, United States, 504-280-7457, unopress.org

Table of Contents

National Parks from North to South

The impressive waterfalls of the Iguazú, the picturesque seascapes around the Nahuel Huapi in northern Patagonia, the iconic volcano Lanín, lined by araucaria forests and the glacier shear edge of the endless ice of the Perito Moreno characterise our landscape imaginaries of Argentina and attract thousands of visitors to their national parks every year. In 2017 alone, 3.7 million tourists visited these parks, a five-fold increase between 1990 and 2017 (Herman 2018).

The genealogy of nature reserves in Argentina is remarkable, being the first national parks to be established in Latin America. The question arises as to why the national park idea was able to establish itself so sustainably in Argentina. The geographical imagination of the nation, which is based on these mostly aesthetically presented natural landscapes and natural wonders in the national parks, was hardly foreseeable at the end of the 19th century. Here, intellectuals who followed Domingo Faustino Sarmiento positioned themselves along the dividing line between civilization and barbarism, the desert or the pampas against urban, European-oriented Argentina. The national parks that emerged at the beginning of the 20th century – above all the National Park of the South established in 1922 (from 1934: Nahuel Huapi) and the planned National Park of the North (from 1934: Iguazú) – had a decisive influence on making peripheral landscapes an integral part of the nation. (Silvestri 2011) With the moral imperative "Becoming acquainted with the fatherland is a duty" (Conocer la patria es un deber) as its motto, the National Park Authority, established in 1934, persuaded the Argentinians to get to know their nation and at the same time to attract foreign tourists to the country.

In the conventional narrative, the establishment of national parks sounds as follows. In 1903, the first National Park was founded at Lake Nahuel Huapi, thanks to a donation by the famous geographer and explorer Francisco P. Moreno. After that, the national parks existed for decades only on paper. The true history of the national parks in Argentina only began with the passing of a national park law and the associated establishment of a national park authority. This story was also stylized into a myth by Exequiel Bustillo, the founding director of the National Park Authority, in his autobiographical writing *El despertar de Bariloche*. But on closer examination, this narrative has many gaps and ignores both the locally and globally entangled dynamics that were of the highest relevance in the Argentine national park debate.

For a critical, transnational interweaving of the genealogy of the national parks in Argentina, three fundamental changes of perspective are therefore undertaken in this book. First, the history of national parks in Argentina is analyzed in a transnational space of entanglement. For semantic reasons alone, a prominent approach usually tells the history of national parks in a national research framework. Parallel to this national narrative, another narrative has also established itself in the inter-American context, which we would like to call an inter-American transfer narrative. The idea of the National Parks is generally regarded as an original U.S.-American invention that is entirely in the spirit of the paradigm of a genuine and unique U.S.-American exceptionalism. And indeed, the first national parks – like the Yellowstone National Park in 1872 – were also established in the USA. From an inter-American perspective, it might be obvious to regard the subsequent establishment of national parks in Latin America as a simple export of the American model.

But both approaches – the national and the inter-American narrative – fall short insofar as they do not focus on the complex transnational circulation of ideas on nature conservation and park concepts, the exchange of experts, and the adaptation of biota and other artefacts. The national parks and protected areas that were established in the Cono Sur at the beginning of the 20th century show, however, that the transfer thesis is too simple. Not all debates can be reduced to unilateral, inter-American flows. Instead, the establishment of national parks is an open process in a transational force field in which there are different influences. These range from French landscape architecture and Belgian park ideas to Prussian sustainable forestry and the German model of natural monuments. Carl Curt Hosseus, one of the lesser known protagonists of the National Park Movement in Argentina, explicitly refers in a 1916 brochure on the National Park of the South, nowadays the Nahuel Huapi National Park, to European experiences from Germany, Belgium, Denmark, France, England, Italy, the Netherlands, Norway, Austria-Hungary, Russia, Sweden and Switzerland. In addition, the USA, Canada, Japan and the Dutch colony Java are mentioned as contributors of key concepts of the national park idea in Argentina (1916: 34-35).

Secondly, this book rejects a linear historiography according to which – starting from the national parks that exist today – a simple stringent narrative is constructed to the origins of these very parks. This linearity is very present in Argentina precisely because of the developmental history told by Bustillo, which finds its crowning conclusion in the own work of the national park legislation and the successful institutionalization of the national park authority. Instead, a rhizomatic historiography is applied, which also traces the parks that have not been realized or have historically disappeared. The traces left in the sources were followed and attempts were made to work out interrelationships. This is a "follow-the-history" that – following on from current debates in socioanthropological-historical transnationalization research (Clifford 1997) – follows people, artefacts, biota and ideas. For the historian, this means not only working with a corpus of sources that is firmly established in an archive, but also chasing traces of the past through a wide variety of archives in his detective pursuit. In such an undertaking, the focus is on supposed secondary figures as well as local actors who implemented the park idea on site and who in part were able to influence the national debate out of their local contexts.

Thirdly, the integral connection between national park and colonialization must be analyzed in order to grasp the significance of the national park in the Argentine history of the late 19th and early 20th centuries. At the end of the 19th century, Argentina had appropriated the peripheral areas in the north and south – some of which were controlled and inhabited by indigenous peoples – and was now faced with the task of sustainable colonization and control of these regions. According to Bustillo, the "National Park proved to be a true instrument of colonization" (Bustillo 1997: 15). This finds expression in the subjugation of indigenous peoples, especially the Mapuche in the south, projects of agricultural colonization and the design of completely new, modern cities as an integral part of the national park regime. Parallel to this, the national parks also proved to be true distribution points for the circulation of neobiota – from red deer and bison to rainbow trout, Douglas fir and wine rose.

With these three shifts in perspective, this book contributes to a better and more complex understanding of national parks without retrospectively projecting the current national park definitions of the International

Union for the Conservation of Nature onto other historical constellations. Instead, it expresses how open and controversial the national park concept was in Argentina between 1903 and 1943. While some scientists concerned with nature conservation reacted with a park concept to the environmental destruction of the pre-anthropocene, other geopolitically or economically oriented actors saw the parks as instruments of territorial control and development engines of international tourism. A field of tension that has lost nothing of its topicality even today.

***

I would like to thank Grace Dolcini for her marvelous translation of this book from German to English. All Spanish, French, and German quotes have been translated into English by the author and translator in order to facilitate a smooth lecture.

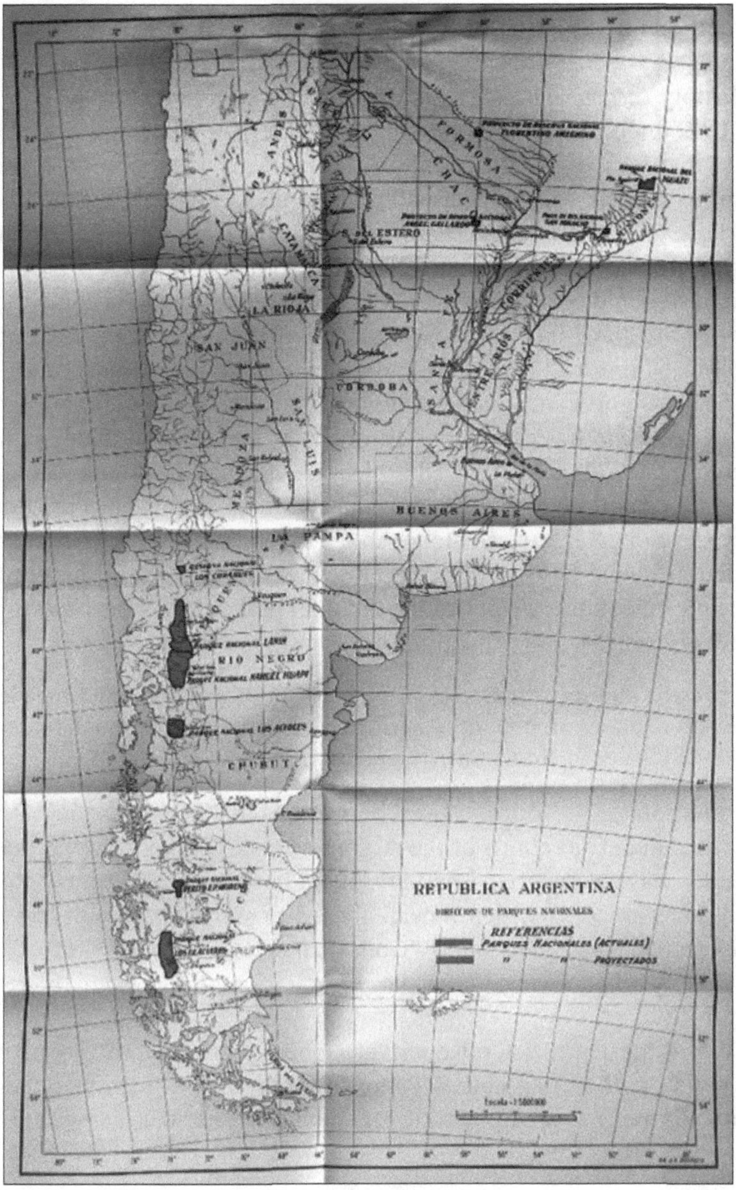

"National Parks in Argentina, 1937", DPN 1937.
Nuevos Parques Nacionales. Buenos Aires.

Iguazú – The National Park of the North

The history of national parks in Argentina began in 1897 in the south of Brazil. Edimundo Barros, commander of the Brazilian military colony Iguazú that was established in 1888, set up a wooden sign on a stately tree on the border with Argentina in the immediate vicinity of the Iguazú waterfalls. The sign read: "Parque Nacional. Março 1897. Edimundo Barros." This was a performative act par excellence as the first one to put up a sign and proclaim the area protected by the state became the inventor of the National Park. But this performative act in the midst of the subtropical rainforest would probably soon have been transported into historical insignificance by termites and bacteria if the German-Argentinean expedition traveler Carlos Burmeister[1] had not stumbled upon the aforementioned sign in June 1899. Burmeister had been commissioned by the Dirección de Agricultura y Ganadería to explore the previously barely developed area of Misiones in the northeastern tip of Argentina – especially for the cultivation of mate tea. Burmeister could hardly hide his astonishment at the sight of the national park sign. In his report, Burmeister argues:

> It seems that this Mr Barros, captain of the Brazilian army, intended to reserve an area around the waterfalls for a national park, much like in the United States in North America. (1899: 22)

This anecdote of the Burmeister report must have been alarming for the geopolitical considerations of the Argentine state. The northern border in particular was characterized by unresolved border demarcations between Brazil and Paraguay, which repeatedly led to skirmishes that were only finally settled with the Laudo Arbitral under the commission of U.S. President Stephen Grover Cleveland in 1895 and contractually recognized in 1898.

While the Burmeister Report brought the concept of the national park – via the Brazilian detour – into the sphere of Argentina's national politics for the first time, the long-standing governor of Misiones (1896-

1 This must be the eldest son of Carl Hermann Conrad Burmeister (1807-1892), a highly respected German natural scientist, who moved to Argentina in 1861 to take over the direction of the Museo Público.

1905), Juan José Lanusse, worked in parallel on regional development through agricultural colonization and above all through tourism. Lanusse himself had visited the waterfalls in 1898 with his family, and many adventure travelers followed in his footsteps, some of whom wrote extensive travel reports illustrated with photographs and drawings. The description of the journey with the Iguazú Falls as a culmination almost became a literary genre in its own right.

In 1901 the first steamboats – España and Iberá – sailed on the Paraná towards Iguazú.[2] In the same year there was the first organized tourist expedition to the Iguazú waterfalls. However, due to the poor traffic infrastructure, only three of the 22 participants could enjoy the magnificent view of the waterfalls. One of the participants, the porteñer aristocrat Victoria Aguirre, then donated 3,000 pesos for the construction of paths to the waterfalls and built one around a landing stage. The latter is considered the origin of Puerto Aguirre (Thays 1912). There were also other investors who rudimentarily expanded the tourism infrastructure, so that in 1922 there was a simple hotel.[3]

This local and regional development dynamic was then combined with the nation-state effort to gain geopolitical control over the peripheral region and to integrate it into the nation via a national park. In order to advance this integration, the Gobernación del Territorio de Misiones and the Ministry of the Interior commissioned the renowned landscape architect and botanist Carlos Thays to travel to Iguazú and explore the tourist and environmental prospects there (Berjman 2002a: 30). The Belgian-French landscape architect came to Córdoba in 1898 to build parks. He then settled in Buenos Aires, where he held the post of Director de Parques y Paseos of the city from 1891 to 1920 and, as founding director, designed the city's Botanical Garden.

On April 6, 1902 Thays and his wife left Buenos Aires for Misiones. The trip on the Rio Paraná led him via Posadas to Puerto Aguirre, the gateway to the Iguazú Falls. He named the lack of infrastructure as a central obstacle to tourism development:

---

2 See Bernárdez 1901 and Pusso 1912.

3 For a detailed discussion of the regional development around the Iguazú cases and the establishment of the Iguazú National Park see Ferrero and Pyke (2016; 2017), Wilde (2007), Freitas (2018) and for the Brazilian side Freitas (2014).

> A major obstacle to the use and popularization of this magnificent subtropical region are the difficult access conditions, which are still protracted and deficient. Not so long ago, it took ten shipping days to reach the Iguazú, a journey almost as long as the journey to Europe!

Upon arriving in Puerto Aguirre, Thays – in coordination with the governor of the Territorio Nacional de Misiones (since 1953 province of Misiones), Lanusse, chose a suitable place for a chalet hotel, in order to promote tourism to the waterfalls, which "even though they do not have as abundant a flow of water as the Niagara Falls, they surpass them by their natural qualities and exceed their most picturesque appearance". (Thays 1902: 44)

EL INGENIERO THAYS INDICANDO LA UBICACIÓN DEL CHALET-HOTEL.

"The engineer Thays determines the location of the chalet hotel" (Thays 1902).

The report in the leading magazine *Caras y Caretas*, richly illustrated with photographs, thus served not only to document the journey, but also to establish the Iguazú waterfalls as a magnet for national and international tourism. The multi-page report had a broad impact, especially the spectacular landscape photographs taken by Frank W. Bicknell, Thays' U.S. companion. Bicknell was appointed Special Agent for Agricultural Research in 1902 by U.S. President Theodore Roosevelt, whose mission was to study Argentine agriculture. Although he himself was involved in livestock and wheat farming in Argentina, he later became known in the USA as a landscape photographer in North Carolina (Bicknell, PC).

In this report, the establishment of a national park is not explicitly mentioned, but Thays presented the Argentine authorities with a detailed plan for the establishment of the national park, which included the development of infrastructure. The Swiss-Argentinean naturalist Eugène John Benjamín Autran, who held the post of Inspector de Agronomía de la Dirección de Agricultura in Buenos Aires from 1908, argued in his contribution to the *Boletín del Ministerio de Agricultura*, "Les Parcs Nationaux Argentins. Le Parc National de l'Iguazú. Le Parc National du Nahuel Huapi et sa florule" in 1907, that Thays was the main author of the plan to establish the National Park Iguazú. Autran notes that after Thays spent "some time in the waterfall area, he designed a comprehensive project that was approved by the government in July and in 1902 was to be implemented" (Autran 1907: 4).

On the basis of these ideas, the engineer Pedro Ezcurra, Undersecretary of State at the Ministry of Agriculture, drew up a plan to implement the Iguazú National Park (Autran 1907: 5). But there were problems with the allocation of land for the park. These were only resolved in 1909 when a national law (the law 6721) was passed to promote Misiones, which explicitly protected the landscape around the Iguazú and authorized the executive office to acquire land for the planned national park by exchange or purchase (Prat 2017). The aim was – according to the law – to establish "a large national park and other works to embellish the immediate surroundings of the large waterfall and the access roads to the waterfalls" (Law 6721, 29.9.1909). The land necessary for this was acquired by the state from the large landowner Gregorio Lezama in a public auction. Lezama did not attach high value to the area and described it soberly as a "jungle area adjacent to some waterfalls" (quoted in Prat 2017). In Buenos Aires, on the other hand, there were intense political discussions about the establishment of the national park. The deputy Candioti, spokesman of the commission for public works in the Chamber of Deputies, considered the establishment of the Iguazú Park a milestone:

> The public land to be acquired by the government will be used for agriculture and the formation of a national park. This is completely new. A park that will cover 10,000 hectares of land near the Iguazú Falls to create a site in the spirit of Yellowstone Park in the United States. (Diario de Sesiones de la Cámara de Diputados, 29.09.1909: 771-772)

In order to define the use of the national park and to advance planning, the expertise of Carlos Thays was again sought. And so, Thays made another trip to Iguazú that resulted in a detailed report, richly illustrated with photos, entitled *Parque Nacional de Iguazú*, which the famous landscape architect presented to the Minister of Agriculture, Adolfo Mujica, in March 1912. In doing so, he revised some of the ideas he had already developed in 1902 for founding the park, the core of which was an urban development project for the village of Iguazú and the military colony (Freitas 2018: 108-110, Berjman 2002a: 30).

Thays also explicitly wrote of the national park project in this report, which he understood to be a similar idea to that of Roosevelt's national parks in the USA. Specifically, Thays draws a comparison to the Niagara Falls, one of the central cosmopolitan tourist destinations. Thays responds to the frequently voiced criticism of tourism and the loss of wilderness at the Niagara Falls, where he counters this assessment with the Iguazú cases, which are still in "an untouched state that Niagara Falls lost many years ago". President Theodore Roosevelt's intervention to protect the Niagara Falls, which was stylized as a heroic deed, could then serve as a model for the – fortunately earlier – protection of the Iguazú Falls:

> Later, President Roosevelt, in view of the fact that the New York State Legislature had granted several industrial companies the exploitation of the hydropower generated by the waterfalls, intervened personally and with all his authority to prevent the complete destruction of the natural beauty of these waterfalls.

The timely intervention in Argentina to protect Iguazú places Thays precisely in this heroic position of preserving wonders of the world for national heritage:

> Fortunately, the Iguazú Falls and the surrounding region are still preserved in their wild and picturesque splendour, and thanks to the timely intervention of the national government, they are protected before it is time only to mourn the destruction that has occurred and try in vain to repair the damage that has been caused. (Thays 1912)

At the same time, Thays was writing Argentina into the emerging global history of national parks. Although he saw the establishment of Yellowstone in 1872 as a striking reference point, he also referred to other initiatives in the USA, Switzerland (Valle Cluoza, Engadina), Australia (Parque Illowa), Germany and England. However, he fails to address the specific approaches of the different national park regimes. Thays himself, however, can hardly be seen as a pioneer for wilderness and conservation as the arguments for nature conservation were directly taken from Roosevelt. However, he becomes more eloquent when it comes to the park's contribution to regional development, the "progreso del territorio". Here, the expansion of the transport infrastructure – from railways to steamboats – is explained to be just as important as the expansion of accommodation facilities. However, the core element of the report was the creation of a new city of Iguazú, with a School of Forestry (Escuela de Selvicultura), a meteorological station, an agricultural experimental farm, experimental fields and a zoological station. These installations alone made up 6,300 of the 25,000-hectare national park in the plan. The remaining 18,700 ha would also be developed with hydroelectric power plants, railways and roads, with a hotel and a casino to be built in the immediate vicinity of the waterfalls. The aesthetic experience of the imposing waterfalls would be guaranteed by a network of footpaths with footbridges, transport bridges and viewing platforms.

In his plans, Thays was always concerned with compensation and "to balance the two concerns presented, namely: aesthetics and industry" – as he pointed out with regard to possible water power projects. However, he also emphasized possible hydropower projects that would not commit the same error as Niagara (Thays 1912 in Berjman 2002: 332). Again, nature conservation itself hardly played a role for him. The central economic sector that promised to combine landscape protection and economic interests was tourism. But Thays did not seem to want to rely on it alone and so the planned railway line served not only to transport tourists, but also to export mate tea and tropical wood:

The new railway line, beyond its strategic importance, allows the extraction of the internal araucaria forests, which are found throughout the central region of Misiones, the simple export of mate tea and the creation of new industrial agricultural centers. (Thays 1912 in Berjman 2002: 325)

"Carlos Thays in front of the Lanusse-Cascade" (Thays 1912).

In addition to this dubious harmonization of landscape aesthetic and economic interests, geopolitical considerations in the border triangle of Iguazú played an important role in the establishment of the national park. Thays points out, for example, that Brazil had been operating a military colony on the banks of the Upper Paraná for many years – which is why a military colony was now to be established on the Argentine side on 1,500 hectares of the planned national park. These plans were very ambitious. The urban vision in particular was in line with a long list of modern planned cities in tropical and subtropical regions, an example of which was the founding of the Brazilian capital Brasilia.

The real implementation of the national park, however, progressed slowly. In 1928, the Argentine government purchased a further 75,000 hectares in the vicinity of the Iguazú Falls. The decisive date that finally catapulted Argentina into the international avant-garde of conservation-oriented countries was November 29, 1934, when Law 12,103 was enacted, which simultaneously established today's Iguazú National Park and Nahuel Huapí National Park. Despite Barros' early efforts, Brazil followed suit five years later and founded the Iguaçu National Park on the Brazilian side of the world-famous waterfalls on October 29, 1939.

Nahuel Huapi – The National Park of the South

A second hotspot for the establishment of national parks in Argentina was located in the remote and barely developed southern part of the country. The region had only been conquered a few years earlier by the indigenous Mapuche population in the *Campaña del Desierto*, and was, like Misiones, also marked by massive geopolitical border disputes. In order to clarify the conflicting territorial claims between Chile and Argentina under international law, a scientific commission was set up to clarify the 4,000 km long border line. Under the mediation of Great Britain, the border treaty still in force today was signed by both countries in 1881.

The central figure for the surveying on the Argentine side was the expert, or *Perito*, Francisco P. Moreno. Moreno knew Patagonia like no other scientist and explorer of his time. Starting in 1875, Moreno travelled and explored Patagonia – which was still controlled by the Mapuche at that time. Emilio Frey, a young Argentinean of Swiss descent and collaborator of Moreno, also attributed the honor of hoisting the first national flag in the region to him on 20 January 1876. As a reward for his services as an expert in the border dispute, Moreno was given a land donation in the eastern part of Lake Nahuel Huapi by the Argentine government. And Moreno for his part donated a section of this area – to be precise, an area of "tres leguas cuadradas" – back to the Argentine state with the condition that a national park be established here (Wakild 2017: 39-42). In a letter dated 3 November 1903[4] to the Minister of Agriculture Wenceslao Escalante, Moreno justified his donation, among other

4 There are uncertainties regarding the exact dating. In an article from the newspaper *La Nación* dated November 8, 1903, the letter was printed by Francisco Moreno and was dated from November 6, 1903. In the archive of Moreno in Bariloche, however, there is a letter dated November 3, 1903. Besides the different dating, there are also minimal but decisive differences in the wording. While the letter from November 3rd uses the designation "parque nacional," in the letter from November 6th, the term "parque natural" is used in the same quotation that I used here in this text. And last but not least, however, the introduction from *La Nación* to the Moreno Letter still speaks of the "formación de un parque nacional."

things, by comparing it with the establishment of national parks in the USA where he writes:

> During the excursions I have made in these years in the south, I admired the extraordinarily beautiful places there. This later motivated me to donate. More than once I have expressed the conviction that the nation must reserve the property for the best benefit of present and future generations, following the example of the United States and other nations that have outstanding national parks. (Letter Moreno to Escalante, 3.11.1903, MP, Colección Moreno, Caja 7)

Moreno's wish was that the region around the Nahuel Huapi "is protected as a public national park" (Letter Moreno to Escalante, 3.11.1903, MP, Colección Moreno, Caja 7). A few months later, on February 1, 1904, the President of the Republic, Julio Argentino Roca, accepted the donation and declared the area a National Park. This donation from Moreno could not have been timed better as on 8 January 1903, Law 4167 had already been enacted in Argentina, which regulated the protection of forests on public lands, as long as no explicit "Ley de Bosques" existed. In Argentina and especially in Patagonia, Moreno's donation is considered in many places as the founding act of national parks. In 1907, the park was extended by a presidential decree from José Figuero Alcorta by 43,000 hectares.

The contemporary sources of that time also show how much the gesture of the perito was appreciated. *El Diario* ran an article titled "The Nahuel Huapi Park [...] An important international work," on February 8, 1904 consisting mainly of fragments of the founding letter of Moreno that emphasized the importance of the national park around Lake Nahuel Huapi. In the concluding paragraph, however, it was also reported that immediately after the legal establishment of the park, the first steps for its implementation were to be initiated, namely by Carlos Thays, whom we already known well. *El Diario* reports, "We would like to point out the importance of Mr. Thays travelling to this place in order to put the noble idea of the former Argentine expert into practice as soon as possible" (El Diario, 8.2.1904).

And Thays then started on his way – albeit some six years later. The government of Roque Saénz Peña (1910-1914), who also promoted a program of national development and colonization with the Minister of Agriculture, Ezequiel Ramos Mexía, commissioned Thays to study the foundation of the park (Thays 1913 in Berjman 2002: 349). Due to the miserable infrastructural connection of Patagonia to the capital Buenos Aires, the journey initially ran by train from Buenos Aires to the west via Mendoza over the Andes pass along the Puente del Inca to the Chilean capital Santiago de Chile. At that time, in 1910, the Transandine railway had just been completed on the occasion of the Exposición Internacional del Centenario. Thus, Thays also recommended this technical masterpiece, which connected Argentina and Chile via a tunnel at an altitude of 3,200 m, for the tourist development of Nahuel Huapi (Thays 1913 in Berjman 2002: 345). Carlos Thays made an intermediate stop in Santiago, where he met Mr. Renner, the director of public parks in Santiago. Here he also made the acquaintance of Federico Albert, a Chilean of German descent, who helped establish the first forest reserves and a general inspection for forest, water, and hunting in 1912. Albert is generally regarded in Chile as the founding father of today's antagonistic areas of national parks and natural heritage on the one hand and modern forestry on the other.

Federico Albert then invited his Belgian-Argentinean colleague to accompany him to Puerto Montt, about 1,600 km south of Santiago (Thays 1913 in Berjman 2002: 346). The two tree and forest lovers must have exchanged intensively, because Albert's voice can be clearly heard through Thays' pen when he formulates his lecture to the Touring Club in France:

> We enter the vast region that used to be populated by trees and is now covered with charred tree trunks that were burned by fire – deliberately, accidentally or criminally. These are the sad remains of the vanished beauty. (Thays 1913 in Berjman 2002: 347)

And Thays continues:

> How many millions of trees have been destroyed by forest fires without any compensation? And when one considers that the natural forest regions of Argentina and Chile unite between the high mountains of the Cordillera: how often did the terrible devastation spread unhindered from one nation to another! (Thays 1913 in Berjman 2002: 347)

In order to counter deforestation, they planned a political-scientific binational congress on forest protection, politically promoted by Dr. Lorenzo Anadón, Argentine ambassador in Chile and enthusiastic environmentalist, as well as Pedro García de la Huerta, president of the Consejo de los Bosques. The aim of this Congreso forestal argentino-chileno was "to reach an international agreement on the protection, conservation and expansion of forests on both sides of the Cordillera" (Thays 1913 in Berjman 2002: 347). On the Argentinean side, Orlando Williams, president of the Sociedad Forestal Argentina, was the driving force behind this congress, which, however, was ultimately realized on a small scale (Tobal 1950: 110-111; 207). Thays and Williams urged the expansion of civil society organizations to protect the trees in the border area. Specifically, the founding of a Sociedad Forestal Chilena, similar to the Argentine society, was discussed (Tobal 1950: 110-111).

Unfortunately, there is no report of Thays' visit to Nahuel Huapi, and since Carlos Thays Junior wrote his dissertation in 1918 at the Faculty of Agriculture at the University of Buenos Aires on the topic "Consideraciones sobre la Formación del parque Nacional del Sud Nahuel Huapi" without mentioning a possible report from his father, it can be assumed that there is no publicly accessible document here.

In contrast to Iguazú, Carlos Thays' visit to Nahuel Huapi for the establishment of the national park there was largely inconsequential. In Bariloche, the cultural capital of the park, the intervention of the Belgian-French landscape architect has largely fallen into oblivion. All the more prominent, however, is the figure of the U.S. geologist Bailey Willis – in addition to the omnipresent Francisco P. Moreno – who, at the same time as Carlos Thays, had worked out a concept paper for the establishment of a national park around Lake Nahuel Huapi in 1913, which we will talk about later. But before this, the contemporary debates on nature conservation are at the center of this analysis.

Love of the Tree: The Sociedad Forestal

> On one occasion I saw two workers in Lezama Park, who had 'nails' in an araucaria to hang a string from. Outraged, I yelled at them to stop their work. As they did not want to obey me, I called a guard to fetch them from the ladder. They then obeyed the order, but when Varela Castex and Francisco Uriburu heard about this incident, they came furiously into the park to tell me that I did not have to interfere in these matters. I then went to Alcobendas, the mayor at the time. Shortly after I was with him, Varela Castex and Uriburu also arrived to document their protest against me. But there in the office Alcobendas wrote a protocol in which he confirmed my attitude and granted me sufficient power to act. My accusers were then pulled away ... At that time the trees were the poor victims of the population. For childlike reasons, thick ombue trees or magnificent araucaria were uprooted or felled. There was no love for the tree. (Thays 1929 in Berjman 2002: 375)

This episode reveals the then director of the Dirección de Paseos of Buenos Aires and founder of the Botanical Garden to a reporter of the magazine *Caras y Caretas* during an interview in 1929. This anecdote expresses a love of the tree that is not only a personal fad of the French landscape architect. Rather, it expresses an attitude of mind shared by many people in Argentina. This is also evident in the lively participation in the public holiday for the tree. The Thays scholar Sonia Berjman also attributes the introduction of the Day of the Tree in Argentina to the Belgian-French tree lover, which was first publicly celebrated on 11 September 1903 on the occasion of the anniversary of Sarmiento's death (Berjman 2002a: 35). This celebration, which was partly accompanied by speeches praising the power of the trees, and partly by public planting of trees, was a cross-class public event in which several thousands of people took part.

At this point, however, it becomes clear that the protection of trees beyond only Carlos Thays was also at the center of the social activities of the educated elite. In 1910, on the instigation of the daily newspaper *La Prensa*, the Sociedad Forestal Argentina was founded and Orlando Williams became its president. Williams was a politician and member of parliament, but also had good contacts in the academic field as an engineer. In addition, he was an avid proponent of tree protection in the feature pages of the Argentine press such as *La Prensa*, *La Nación*, *La*

Razón, *Anales de la Sociedad Rural Argentina*, *Revista de la Bolsa de Cereales*, as well as in his own publications *Boletín* and *Anales* of the Sociedad Forestal. Williams had already been pushing ahead with a law as early as 1902 which, among other things, provided for the establishment of a tree nursery in the Delta de Paraná, explicitly inspired by the forestry policy of the USA and its appreciation for trees (Tobal 1950: 36-38).

One of the explicit aims of the Sociedad Forestal, as stated in the second article of its statutes, is "to promote the conservation and replanting of natural forests" (quoted from Tobal 1950: 94). The Sociedad worked towards the adoption of forestry legislation, promoted scientific work, and established public civil society activities, such as the Day of the Tree. Despite a variety of activities, however, there was no significant direct and sustainable intervention in the debate about the establishment of national parks, but there were good friendly connections. For example, Carlos Thays added a personal dedication to his book on the botanical garden in Buenos Aires that he donated to Williams, and in which he praises the tree lover as a "constant friend and defender of trees" (Thays quoted from Tobal 1950: 53).

The Sociedad Forestal was closely integrated into the club network of the aristocracy in Buenos Aires, which was centered around the Jockey Club. Williams was a member, as were Estanislao Zeballos and Adrián Ruiz Moreno, and aristocrats such as Aarón de Anchorena and Ortiz Basualdo, as well as the prominent natural scientists Ángel Gallardo and Carlos Thays (Tobal 1950: 469). Many of these names would continue to play an important role in the founding process of the national parks in Argentina.

Beyond the charitable orientation, which also corresponded to the spirit of the time of aristocratic public engagement, tree protection also had a political orientation. Taking a pan-American perspective, Williams referred to the immense destruction of nature in the course of progressive modernization and incipient industrialization:

> At present, vast forests are being destroyed in America, and nobody cares about their replanting. The brutal destruction of what is physically the Earth's lungs deserves our youthful benevolence, the very purpose of 'forestry.' (Williams quoted after Tobal 1950: 194)

It is noteworthy that he was already introducing the trope of the jungle as the lung of the earth – a picture that still characterizes the environmental policy debate of protection of the Amazon rainforest today.

Beyond this environmental policy inventory, Williams also referred to the close articulation of extractivist plundering of nature and colonialism, which continues to this day. He describes the forestry industry of his time as "an industry like that of the Indians, when they traded gold for eye-catching glass bead necklaces. Today we supply the gold of our forests, for the beautiful coloured glass, the tax rights". With supposedly eco-apocalyptic echoes Williams then expresses a prophetic warning, which in the current era of the Anthropocene can by no means be regarded as exaggerated:

> Prepare for droughts, plagues and fears for the future. The disorder of nature will one day impose the replacement of what is plundered today as loot of conquest without foresight and without borders. (Williams quoted after Tobal 1950: 194-195)

The social significance of the tree-friendly movement would continue years later. During the Fiesta del Árbol in 1916, Dr. Lorenzo Anadón, the former Argentine ambassador to Chile, gave a speech to a mass audience in Buenos Aires, and even the President of the Republic did not miss this event. There he spoke again of the history of the holiday as well as its current significance:

> The Festival of the Tree begun in the United States to compensate for the gigantic destruction of the primeval forests that populated this land no more than fifty years ago could and should be immediately accepted by us. (Andón 1916 quoted in Tobal 1950: 401)

As already mentioned, the establishment of national parks is one of the goals listed in the constitution of the Sociedad Forestal to stop the destruction of forests. The Minister of Agriculture, Eleodoro Lobos, also emphasized the success of the Argentine government in establishing national parks in Iguazú, Nahuel Huapi, Tierra del Fuego, Chaco and Puno Hacho during his speech at the Day of the Tree in 1911. The Minister Lobos placed Argentina in a global debate context on nature conservation, in which he also emphasized the country's participation in the VIII International Congress of Zoology in Graz, which ultimately led to the international declaration on nature conservation that was carried out in 1913 (Tobal 1950: 297-301).

Similarly, Lorenzo Andón proclaimed in his 1916 Tree Day speech that the creation of national parks were a success of the Sociedad Forestal and its prominent member Carlos Thays:

> Within the framework of these ideas, the establishment of two national parks was planned by the reputable Vice President of the Forestry Society, Mr. Carlos Thays: The first at the shore of Lake Na-huel Huapi and the second at the foot of the Iguazú Falls. The United States and then Canada are the countries that have initiated the creation of these new museums or nature reserves, designed to preserve in large spaces, usually thousands of square kilometers, all animal and plant species, untouched forests, lakes, mountains, picturesque landscapes, in short, everything that represents native nature and that has not yet been trampled by the industrial factory. (Andón 1916 quoted in Tobal 1950: 404)

In addition to these pan-American and cosmopolitan networks, tree protection also has a deeply Argentinean dimension, which is closely linked to national natural myths. Scientifically, the pampas can be defined as a subtropical grass steppe in south-eastern Latin America. In the national imagination, on the other hand, the pampa is regarded as the epitome of barbarism and an obstacle for civilized progress. In this respect, trees in Argentina have also been seen as a natural symbol of civilization. Nature-spatial civilization – in the sense of hygienic improvement and economic utilization – can thus be promoted by a targeted biopolitics of orderly foresting.

1913 – A Climax of the National Park Debate: From Carlos Thays to Bailey Willis

> By 1913 the cult of the tree had spread so far that there was not one city of the Republic that would not have followed this call. And there were so many cities that adhered to this ritual that on a demonstrative map drawn by the engineer Ruiz Moreno, on which the places of the tree festivals were marked, a complete, almost absolute participation from one end to the other of the four cardinal points could be seen. (Tobal 1950: 289)

This contemporary description, which characterizes the year 1913 as the high point of the nature conservation debate, can be confirmed historically. And beyond nature conservation, the year 1913 can be regarded as an epochal change. The publisher, S. Fischer, thus advertises the historical novel *1913* by Florian Illies on the cover text with a striking definition of the year 1913. This is regarded as a unique year "in which the long 19th collides with the short 20th century of wars and extremes" (Illies 2012). Although this reading is Europe-centered in view of the outbreak of the First World War, the effects had also been clearly perceptible in Latin America (Rinke 2015).

Around this year, leading internationally renowned scientists such as the French-Belgian landscape architect Carlos Thays, the U.S. geologist Bailey Willis, the German botanist Carl Curt Hosseus and the Argentine politician Carlos Gallardo presented visionary plans for the establishment of national parks in Argentina. With his lecture "Les forêts naturelles de la République Argentine" which Carlos Thays gave in 1913 on the occasion of the Congrès Forestier International in Paris and at the Touring Club, he set a milestone in the national park debate. Thays took part in these congresses in his capacity as Vice President of the Sociedad Forestal. He thus made the Argentine national parks known to an international audience in a single stroke, as Paris was still regarded as the cosmopolitan capital of Old Europe. In addition to the worldwide circulation, Thays' speech was also distributed widely in Argentina in the form of a brochure and was able to further anchor the national parks in the collective soul of the nation in the Argentine public. Previous studies on the establishment of national parks in Argentina have always focused on only one region at a time. Thays himself had prepared a study on "El Parque Nacional de Iguazú", while Willis concentrated on northern Patagonia and the Nahuel Huapi. In his 1913 lecture, however, Thays presents the Argentinean parks from north to south. The impact of the lecture, however, is not least due to the fact that it was richly illustrated with 220 projected photographs. The idea of this innovative distribution via photographs goes back to Orlando Williams, who also provided the necessary equipment for the Sociedad Forestal (Tobal 1950: 256, 264). The aim was to inspire the public with the visual impressions of trees; "the cult of the tree 'should' penetrate through the eyes" (Tobal 1950: 264). The big performance then took place at the International Forestry Congress 1913 in Paris, where Thays participated together with Miguel Angel Tobal as delegate of the Sociedad Forestal. The success was overwhelming:

> Mr. Thays gave a conference in Paris on 'National Parks' with slide projections, with which he earned the applause of the Assembly and the congratulations of the Argentine Minister in Paris, Dr. Enrique Larreta, who attended the event. (Tobal 1950: 264)

Carlos Thays donated a complete set of photographs including the text of the lecture to the President of the Assembly, Senator Ballif, who was also President of the French Touring Club. In France, the lecture provoked storms of enthusiasm, while at the same time praising the Argentine government as a worldwide pioneer in nature conservation. The President of the Forestry Congress and President of the Comité de los Sitios y Monumentos pintorescos in France congratulated the "Argentine government in showing us the way to go in creating national parks in France, and we can draw inspiration from the examples of this friendly republic" (Deffort in Berjman 2002: 361). The Sociedad's visual material was also presented in Belgium at the Ghent exhibition and in 1915 at the universal exhibition of San Francisco in California.

At the same time, there were other efforts from Carlos Thays' circle to promote the establishment of national parks in Argentina. The Thays student Benito Carrasco, who later became an engineer at the Universidad de Buenos Aires, was also part of this group. Here, Carrasco was the supervisor of Carlos León Thays, the eldest son of the renowned landscape architect. Carrasco was also a member of the board of directors of Sociedad Forestal Argentina and, together with Carlos Thays, was the editor of the publication series *Anales* and *Boletín* of the Sociedad Forestal (Tobal 1950: 237). On October 2, 1912, the daily newspaper *La Nación* published an article in which a study by Benito Carrasco was presented. In this study, Carrasco argues that it is necessary "to withdraw large areas of land from speculation to turn them into national parks, as has been happening for some time in the United States". In order to drive this process forward, Carrasco proposed the creation of a commission to identify the best places to set up national parks. Its aim would be, "to preserve the galas of national heritage for generations to come" (*La Nación*, "La Heredad Territorial. Reservas Nacionales", 2.10.1912). To justify this, Carrasco proposed a patrimonial discourse that encompassed natural and cultural heritage stating, "[i]t is a matter of setting limits to the destruction of our forests and also of leaving the historical ruins and natural monuments undamaged" (*La Nación*, "Parques Nacionales en los Estados Unidos: necesidad de imitarlos", 7.5.1912). The headline in *La Nación* explicitly refers to the model of the USA: "National parks in the United States: The need to imitate them". Carrasco himself was familiar with the debates there, as he had conducted an educational trip to the USA and Europe between 1906 and 1908.

Under the auspices of the Ministry of the Interior, a conference of governors and senior officials was also held in 1913 to discuss a project presented by the engineer Carlos Gallardo, governor of the province of Rio Negro, to protect the country's natural heritage. Specifically, Gallardo called for "the expropriation of land around the Iguazú Falls, the lakes Nahuel Huapi and Pellegrini, and the glaciers of Tierra del Fuego to create parks modeled after Yellowstone in the United States" (*La Nación*, "Las bellezas naturales", 21.5.1913). Carlos Gallardo refers again explicitly to the national park project already developed by Thays at the Iguazú Falls, intended to be universally applied. The project proposed specific sites for the establishment of further nature reserves beyond the two emblematic parks, such as the Sierra de Cordoba, the Valle de Singuil in Catamarca, the Sierra de la Ventana in Buenos Aires; the Valle de Zonda in San Juan, the Quebrada de Lules in Tucumán, or areas that were still used for military purposes at that time, such as Plumerillo in Mendoza, Bella Vista in Salta, Campo de Mayo in Buenos Aires (*La Nación*, "Las bellezas naturales", 21.5.1913).

Carlos Gallardo, governor of the Territorio Nacional de Rio Negro from 1906 to 1913, by virtue of his career and habitual tendencies was predisposed to advance the national park idea. As a botanist, engineer and politician, he dealt academically and politically with the less developed, peripheral parts of the country in Misiones and Patagonia, as well as later with the indigenous peoples of the extreme south, such as the Selknam and Ona. He knew Europe, North America, Africa and Asia from educational trips and was also familiar with North America as the Argentinean representative for the 1876 Philadelphia World Exposition on the occasion of the 100th anniversary of the founding of the USA. Through this contact, the initiative to promote the establishment of national parks through the governors was presumably also inspired by the USA. U.S. President Roosevelt had invited the governors of the various states of the United States to the White House to discuss the expansion of nature conservation (Brinkley 2010: 767, 770). On the basis of this meeting, far-reaching successes in nature conservation were achieved in the USA.

In addition to the initiatives around the Ministry of Agriculture and around Carlos Thays, who had been involved in establishing national parks since 1902, a further field of force developed which promoted the debates on national parks, translated them into concrete policy proposals and also disseminated them in the media. The central figure here was the U.S. geologist Bailey Willis, who had a cosmopolitan career as a scientist. He studied in Germany as a teenager and after graduating in mining engineering, Willis worked from 1881 to 1884 as a surveyor for the Northern Pacific Railroad and subsequently in the United States Geological Survey. From 1895 to 1902, he also taught geology at the renowned Johns Hopkins University. In 1903, he accepted an invitation from the newly founded Carnegie Institute to lead an expedition to northern China and in 1910, Willis was commissioned by the Argentine government to carry out hydrological studies in Patagonia, for which he travelled and researched between 1911 and 1914. In addition to the basic work, the main aim in Argentina was to explore the construction of a railway line to open up the region around the Nahuel Huapi. This commission was appointed by the Minister of Public Works, Ezequiel Ramos Mexía, a committed protagonist of the colonization of Patagonia and created in the spirit of Francisco Moreno's vision of integrating Patagonia into the nation.

The geologist Willis was good friends with the geographer Moreno. And the two scientists often talked extensively about the establishment of the National Park at Nahuel Huapi, as Willis reports, "During my visits to Buenos Aires he and I discussed the campaigns like comrades-in-arms and the realization of the purpose of the National Park was a frequent subject of concern" (Willis 1947: 96). Before his arrival in Argentina, Willis had already heard of Moreno's idea of the National Park from the English traveler and later British ambassador to Washington, James Bryce. But even Bailey Willis himself was no blank page in the field of setting up national parks. He had already been involved in the establishment of Mount Rainier National Park in the USA (1947: 97). Willis compares both parks as the inter-American northern and southern poles of the western hemisphere having much in common:

> The two objectives had one great central feature in common. Mount Tronador dominates the southern park as Mount Rainier rises grandly in the northern. Both are volcanic cones, both are ice-clad. And if the northern forest more profoundly convey the impression of cathedral aisles, there is nothing in the north to compare with the gran fiords of Nahuel Huapí. The two areas are equally appropriate and suited to the end that they be dedicated to the nation in perpetuity. (Willis 1947: 97)

Willis was also closely and amicably connected in the other networks of the scientific discussion on nature conservation. He had many conversations with Gifford Pinchot, the father of sustainable forestry in the USA:

> In Pinchot I found a comrade who sympathized with my indignation, and where I had felt helpless to better the conditions, he was impelled to fight them. I recall an afternoon when we two enthusiasts discussed the situation in the quiet of the Cosmos Club. He talked of the cultivated forests of Europe, plated by one generation to be harvested by another, I of the virgin wilderness. In think his mind was already made up. He accepted the challenge of thoughtless greed and started the movement for conservation which he carried forward as long as he lived. The nation owes him a debt that only future generations may fairly evaluate. (Willis 1947: 34)

Deeply influenced by these conversations, Willis was more inspired by Gifford Pinchot's idea of conservation than by John Muir's approach to preservation. In this sense, he saw no fundamental contradiction between conservation and economic use of natural resources:

> Many believe that exploitation and conservation are contradictory words, the former being seen as a significant destructive use for immediate gain

> and the latter as a representative of the idea of conservation for future use. Such a view has been often proven wrong. The exploitation of natural wealth with due care to avoid waste and promote post-harvest regeneration is conservation of nature. Conservation means using the harvest as it ripens, i.e. it should not be used or destroyed before it has reached maturity. (Willis 1988: 9)

On the basis of these general considerations, Willis writes of the specific problems of Argentina, where he identifies two fundamentally different constellations with regard to the protection of forests stating, "Argentina's forests are limited, and they can be divided into two different classes, namely the tropical and the temperate Cordillera. These differ not only in the tree species, but also in their use" (1988: 10). Contrastingly, the forests in the northern Andes were still largely state-owned and could be managed by a state forestry service. Willis' main problem was uncontrolled forest fires. In the first half of the 20th century, forest fires that were deliberately or accidentally caused posed such a serious threat to Patagonia's forest resources that they coined the bon mot of the time as: "Buenos Aires dances while the Cordillera burns" (Willis 1988: 388). Willis himself named nomadic livestock farmers – probably Mapuche – coming mainly from Chile as the culprits, who burned down the undergrowth in order to sprout new vegetation for the pasture. In order to prevent this, Willis demanded police surveillance (Willis 1988: 11, 170 ff.). This police intervention was to be flanked by forestry regulation, which – in the sense of Pinchot – would culminate in state-regulated scientific forestry.

With regards to sustainable forestry, Willis was therefore quite interested in the use of Argentinean woods. He sent wood samples from northern Patagonia to the Forest Products Laboratory of the Servicio Forestal in Madison, Wisconsin. But the economic exploitation of forest resources in Patagonia was not limitless and certainly not a new El Dorado. This was how the U.S.-American expert Henry Gurfan, who compared the Argentinean woods with North American ones, worded the planned use in his final report of April 8 in 1913, where he concluded:

> As far as we know about the properties of these special Argentinean woods, none of them are as promising in the quality of new cellulose woods as many North American woods, which are currently used rarely or not at all for this purpose. (Willis 1988: 447)

Overall, the economic calculation of the Patagonian primeval forests was therefore sobering. The quality of the wood could not compete with that of North America, and the primeval forests of the Andes were too old for forestry (1988: 384-385). Accordingly, Willis came to the conclusion that "the amount of wood that can be extracted from forests in their current condition is less than is commonly believed" (Willis 1988: 385). In view of the low profitability, he saw the benefits of forests primarily for their role in fragile ecosystems, and more importantly in the regulation of water. Willis considered the latter to be the central resource of the region, especially with regard to its use for hydroenergy. And so, he warned against the thoughtless destruction of forests stating, "If the forest is destroyed, this benefactor will become a destroyer of farms and communities. Europe and Asia present us with countless examples whose teachings should be taken into account" (Willis 1988: 387).

Willis also called for comprehensive state intervention in the enforcement of the national park at Lake Nahuel Huapi. He was commissioned by the Director General de Agricultura, Dr. Julio López Mañán, to prepare a study on the Nahuel Huapi National Park. This study first appeared in 1913 in a brochure of the Dirección de Agricultura y Defensa Agrícola and was later reprinted in his book *El Norte de la Patagonia* with the title "Parque Nacional del Sud". Willis had probably discussed these ideas intensively with Moreno and also with Emilio Frey.

The influence of the Swiss-Argentinean surveyor in these debates should not be underestimated. Frey had already taken part in the expeditions of Francisco Pascasio Moreno as a young man and was considered one of the best contemporary connoisseurs of Patagonia. Frey, who settled in Bariloche, became the central figure of public life here. He also remained attached to the national park idea throughout his life. Accordingly, he became the first director of the later National Park of the South (Polgar 1964). Willis appreciated the work of his first employee so much that he renamed the lake Las Ranas, which he had 'discovered', in Lago Frey.

In his study, Willis not only investigated the area of the actual park donated by Francisco Moreno, but also explicitly included the surrounding public lands, as well as the few plots of the zone occupied by private individuals in the study (1988: 427-428). He initially argues for the park in terms of landscape aesthetics stating that the "Lake Nahuel Huapí is the main feature of the national park" (1988: 428), which does not have to fear any comparison with the scenic beauties of Switzerland, Italy, and the USA. However, he also believes that it is difficult to determine the boundaries of the park, as there are many possibilities for tourism writing that "there are also other uses in the region and so the area of the park must be appropriately restricted". And so, he comes to the following proposal of the delimitation of the park:

> Therefore, an extension equivalent to half of the Yellowstone Park in the United States, i.e. 440 miles, is proposed. The northern border includes the lakes Villarino, Falkner and Traful, the eastern border must coincide with the Limay river valley and the continental separation between Atlantic and Pacific in the south of Lake Nahuel Huapí. The rivers Villegas and Manso can be considered as borders in the south, including the beautiful lakes Steffen, Martín, Mascardi, Hess and Fonck and other small lakes in the southern part of the park. To the west of the park lies Chile and we would have to wait until the Argentinean park connects with the Chilean park on the adjacent mountain ranges. Within the given area, possibilities can be found for the development of all the activities that should be set up in a park dedicated to the welfare of the nation. (Willis 1988: 429; Willis 1913)

In his study, the U.S. geologist also explicitly dealt with the Gretchen question where he asks: "What is a National Park?" He then lists a common prejudice as a rhetorical question: "A wild region intended for occasional hunters or mountaineers who want to face the difficulties of deserted mountains?" And then denies it decisively stating:

> A national park is an area reserved by the state for leisure and for the well-being of the entire population. (1988: 430; 1913)

And then explains further that

> [i]t shall be made available so that no individual may hinder others from enjoying it and so that everything that might benefit man is preserved in its natural state. (1988: 430; 1913)

According to this universally sounding definition, however, Willis then makes a remarkable turn and contextualizes the park model, where he writes. "Different nations have different customs. A national park must be developed according to those of the nation to which it belongs". He then settles on a universalistic-anthropocentric justification for national parks, giving nature a sublime character:

> We can only say that everyone enjoys the beauty of the crystal-clear water and the shady forests and experiences a delight at the sight of mighty cliffs and snow-covered hills. Hence, a region equipped with such nature is obviously destined for a national park. (1988: 430; 1913)

In this sense – starting from the beauty of the landscape in the eye of the Western observer – he compares the park foundations in the USA with those in Argentina stating:

> We have in the United States the parks Yellowstone, Yosemite Valley, Mount Rainier and Glaciers, with some other smaller ones, and so Argentina should have the National Park of the North, around the waterfalls of Iguazú, and the National Park of the South with the Nahuel Huapi Lake and the Tronador. (1988: 430-431; 1913)

As for the concrete design of the park between landscape aesthetics and economic development, Willis then sees Argentine society, mediated through political representatives, as having a duty:

> The development of the park depends on the wishes of the Argentinean population, who can create the conditions suitable for them through the two houses. (1988: 431; 1913)

In the following years, it was to become clear how this Argentine society, which was currently undergoing radical change, should deal with this task. In 1913 the signs were initially not bad for a quantum leap for the establishment of nature reserves. After the aforementioned meeting of governors and other high-ranking politicians for the implementation of the meanwhile ten-year-old national park idea, the media debate on this topic picked up speed again in 1914. On 17 July, *La Nación* published a full-page, richly illustrated and ambitiously laid out contribution to the Nahuel Huapi National Park. The newspaper announced a political breakthrough:

> It seems that the idea of enhancing the natural beauties of the lakes and Andean cordilleras region, especially on the Nahuel Huapi, which had long been bubbling over in the spheres of government, has now entered a

new phase. (*La Nación*, "El parque nacional de Nahuel Huapi", 17.7.1914)

The main problem for the previous establishment was the complex land question, since settlers had long settled in the area of the national park. According to *La Nación*, this problem could now be regulated by the declaration of all unoccupied lands as inalienable public property and by restrictive rules of use for the already occupied parcels.

EL PARQUE NACIONAL DE NAHUEL HUAPI

"Nahuel Huapi National Park", *La Nación*, 17.7.1914.

These newspaper articles may well have been inspired by a project that Francisco Moreno and Bailey Willis had begun in 1913. Both were working together on a book about the National Park, of which Willis wrote "contributed a number of photographs that brought the beauties of the sceneries vividly to his [Moreno's] mind" (Willis 1947: 135).

In addition to the media coverage, the U.S.-American geologist was also politically active as a lobbyist for the National Park of the South. In 1913 Willis presented his project for the organization of the national park at Lake Nahuel Huapi to the Ministry of Agriculture in a 14-page report plus maps. This presentation was flanked by a public exhibition of a "beautiful series of panoramic shots" (*Physis* 1917: 82) in the entrance hall of the Ministry of Agriculture. Specifically, Willis drew up a development plan for the park, which provided for the expansion of the transport infrastructure on water and land, the construction of hotel facilities, support programs for the construction of holiday homes "for the rich people," the granting of long-term land use titles, and the construction of simple holiday homes and campsites for the popular classes (1988: 431). Tourism hopes in Bariloche also seemed to be realized with the expansion of the transport infrastructure. The railway connection was under construction and in 1913 the first automobile, a Mercedes, arrived in Bariloche.

Through these initiatives to establish national parks in Argentina, further momentum was generated which contributed to the intensification of the debate. In 1913, Theodore Roosevelt, former U.S. president and pioneer of nature conservation, visited Nahuel Huapi National Park, which was in the process of being established. In his own travel reports, Roosevelt made no direct reference to the founding of the National Park, but in various talks he would certainly have strengthened the Argentine pioneers of conservation. Roosevelt personally exchanged ideas with Francisco Pascasio Moreno, the elderly initiator of the national park idea in northern Patagonia (Zusman 2010: 6). Despite his old age, Moreno did not miss the opportunity to accompany the illustrious guest with his horse:

> Doctor Francisco P Moreno was in excellent condition and despite his advanced age he rode excellently and galloped more surefootedly than many others present. (Letter Primo Capraro to Emilio Frey, 30.11.1913, MP, Colección Frey, Bibliorato 4; spelling mistake in the original)

For Moreno, this would be his last trip to Patagonia during his lifetime. He died on 22 November 1919 in Buenos Aires. In 1944 his mortal remains were transferred to the island of Centinela in Lake Nahuel Huapi, where they found their last home in a solemn ceremony.

Roosevelt used the trip to Northern Patagonia for – as we would say today – a trekking tour. From Puerto Varas, Chile, he approached the Cordilleras by steamboat, which he crossed by horse over the Paso Pérez Rosales to Bariloche. He compared northern Patagonia with the U.S.-American West and states: "Bariloche is a real city of the frontier" (Roosevelt 1914a). In doing so, as in the Wild West, the confrontation with the indigenous population also plays a central role. But the descriptions of the landscape and the appreciation of the National Park's founding are also essential in Roosevelt's travel report. As quoted in a newspaper article from Francisco P. Moreno's private archive, which made a correction:

> We had been through a stretch of scenery as lovely as can be found anywhere in the world – a stretch that in parts suggested the Swiss lakes and mountains, and in other parts Yellowstone Park or the Yosemite or the mountains near Puget Sound. In a couple of years, the Argentines will have pushed their railway system to Bariloche, and then all tourists who come to South America should make a point of visiting this wonderfully beautiful region. Doubtless in the end it will be developed for travelers as much as other regions of scenic attraction are developed. Thanks to Doctor Moreno, the Argentine end of it is already a national park; I trust the Chilean end soon will be. (1916: 141-142)

A comparison of the Nahuel Huapi region with the U.S. icons of the national parks has often been made, but according to Theodore Roosevelt, this comparison is tantamount to a noble achievement for the region. In addition, Roosevelt was so well informed about the regional development plans, that the Willis Plan, which he also worked out with Moreno, was implicitly referred to here.

Beyond Moreno, Roosevelt's visit was also formative for other people from the national park movement. The former President of the United States also came into contact with Emilio Frey, and the National Park advocate and Director of National Territories Isidoro Ruiz Moreno was part of Roosevelt's tour guide.

Thus, in the decisive year of 1913, there were two elaborated but conceptually different drafts for the institutionalization of the National Parks, which were waiting to be implemented. From the environment of

Carlos Thays, Benito Carrasco and the Friends of the Tree, a national park concept based on landscape aesthetics as well as natural and cultural heritage was developed that is anchored in the national-patrimonial discourse that began in the 19th century beyond the U.S.-American and, above all, European debates on natural monuments. This is supported by the utilitarian and pragmatic concept of Bailey Willis and Francisco Moreno and Emilio Frey, who follow on from the U.S. discourses and focus on reconciling the protection of monumental landscapes with the economic development of peripheral regions.

Yet 1913 proved to be a year of crisis despite the promising pledges regarding the concrete political implementation of the national parks. Willis himself blamed the government crisis in Argentina in 1913 for the failure or stagnation of the national park idea. In his view, the bureaucratic class concerned with national integration was now replaced by a group in political power turned toward big business (Willis 1947: 152). In addition, there were the distortions brought about by World War I, which made Willis laconically state with regard to the hydrological commission he headed: "The Comisión de Estudios Hidrológicos became a casualty of the war" (Willis 1947: 149). Altogether, the efforts of the years 1913 and 1914 did not find a consistent and timely implementation.

This abrupt paralysis of the circulation of nature conservation discourses and practices was also evident on a global scale. Intensive efforts to protect nature and establish nature reserves were closely intertwined with a global dynamic that culminated in 1913. In 1909, European naturalists and environmentalists met in Paris at the Congrès International de Protection des Paysages, where the creation of an international organization for environmental protection was being considered (McCormick 1991: 22-23). This idea was taken up by the Swiss zoologist Paul Sarasin, founder of the first national park in Switzerland, and discussed at the International Conference for the Protection of Nature in Bern in 1913 (International Office for the Protection of Nature 1931: 5). At this meeting of delegates for the World Commission on Nature Conservation, Sarasin (1914) presented his influential, programmatic essay "On the tasks of world nature conservation," which clearly focused on the protection of fauna. From this perspective, the 62-page essay fails to mention the efforts made since the beginning of the century to establish nature reserves in the Cono Sur. This conference was attended by several European countries such as Belgium, France, Netherlands, Germany, Norway, Sweden, Russia, Romania, England, as well as non-European countries like Japan, Australia, the USA and Argentina. The USA was also represented by Kermit Roosevelt, the son of the U.S.-American ex-president and conservationist Theodore Roosevelt

In the run-up to the event, Argentine participation was used by the government to present itself on the international stage. In his speech on the occasion of the first national day of the tree in Argentina in 1911, Agriculture Minister Lobos referred to the successful establishment of national parks "Iguazú, Nahuel-Huapí, Tierra del Fuego, Chaco and Pino Hachado" and that "the nation has joined the Temporary Committee on Global Nature Conservation" (Tobal 1950: 299-300). The Argentine and U.S. delegates sat together at the conference, but – as U.S. historian Ian Tyrell speculates – "it is unlikely they would have found the proceedings entirely satisfactory, since the conference focused on zoological and proto-ecological issues of species depletion rather than national resource allocation as a whole" (Tyrell 2015: 232). Consequently, the decision to found the international organization was ultimately only signed by 17 European states, but not by the USA and Argentina; and overall, the organizational process – probably also due to the turmoil of the First World War – remained in its infancy.

Scientific Highlights and Political Stagnation: Carl Curt Hosseus, Lucién Hauman and the Sociedad Argentina de Ciencias Naturales

Despite the political blockade when implementing the national parks, there was still a continuous interest in nature conservation in civil society and the natural sciences even during the First World War. In 1916, on the occasion of the Day of the Tree, Lorenzo Andón emphasized the efforts of Carlos Thays to establish national parks at the Iguazú and Nahuel Huapi in front of the President of the Republic and a broad public. According to the speaker, these served to protect flora and fauna, and the parks also preserved the "magnificent natural wonder" as they promoted the economic development of the regions (quoted after Tobal 1950: 404-405).

But the success of the Sociedad Forestal was not limited to Sunday speeches. In 1915, the Sociedad Forestal, together with biologists such as Carlos Spegazzini, worked towards the implementation of a *Ley de Bosques y Yerbales*, which, however, was not approved. As part of this legislative initiative, a first nationwide forest survey was carried out, according to which 106,888,400 hectares of the land were covered with forest. After this initiative, however, the importance of the Sociedad Forestal, which had been founded in 1910, declined.

Instead, two new organizations were to emerge in the 1910s from the particularly intense debate on natural science and nature conservation. The Sociedad Amigos del Árbol de la República Argentina (Society of the Friends of the Tree), founded among others by the former member of the Sociedad Forestal Adrían Ruiz Moreno, promoted tree protection as a civil society organization (Tobal 1950: 526). Parallelly, the Sociedad Argentina de Ciencias Naturales emerged as a new relevant actor with a scientific orientation. The statutes for the foundation of the Society had already been drawn up in 1911, but it was not until 6 January 1916 that the Argentine government officially recognized the Society by decree. This foundation was accompanied by the formation of the Academia Argentina de Ciencias Naturales (Argentinean Academy of Natural Sciences) in which renowned natural scientists such as Carl Curt Hosseus and Lucién Hauman participated, and who also played a leading role in the national park debate (Hosseus 2004: 14).

The fourth objective of the statutes of this society is particularly relevant for the question we are pursuing here about the history of the origin of the national parks in Argentina:

> One of the society's main objectives is to promote the establishment of nature parks and the adoption of measures to protect wild fauna (especially vertebrates) and to disseminate knowledge about the benefits of these issues. (Extracto de las Bases de la Sociedad Argentina de Ciencias Naturales in *Physis* 1915)

The latter, the dissemination of ideas on nature conservation, took place primarily through the journal *Physis* published by the society. In the very first volume of *Physis*, published in 1913, the president of the Sociedad Argentina de Ciencias Naturales, Martín Doello Jurado, wrote an essay on the creation of a national park near the urban agglomeration of Buenos Aires (Doello Jurado 1913). In the second volume of *Physis*, covering the years 1915 to 1917, an overview can be found in which the Argentine nature parks are firmly anchored in the international debate on the creation of nature reserves.

> The attractiveness and usefulness of untouched areas, the so-called 'nature parks', is of great concern to governments, societies and individuals in a number of countries, as is the protection of wild birds and mammals. (*Physis* 1917: 80)

The aesthetic and scientific reasons for the establishment of nature reserves are highlighted. The authors adopted the new debate led by Hugo Wilhelm Conwentz on the establishment of 'land reservations' in remote areas such as the polar regions threatened by whaling, as well as his approach of 'preserving natural monuments' (originally named in German: *Naturdenkmalpflege*). Subsequently, the U.S-American debate, which was conducted in the leading scientific journals *Nature* and *Science*, would be presented. Here, too, the protection of glacial regions – according to Norwegian Spitsbergen – was at the center, as suggested by Prof. Greenville. The establishment of new national parks in the USA was also presented and discussed, including publications on existing parks such as Glacier National Park (*Physis* 1915-1917: 208). The U.S.-American icon of the national park movement, the recently deceased John Muir, was particularly honored. His book *Our National Parks* is seen as the starting point and milestone of an international park movement.

But this emergent transatlantic conservation movement – as the anonymous authors criticize – bears only embryonic traits in Latin America in general and in Argentina in particular:

> This spirit stands in remarkable contrast to the carelessness or contempt that Latin America has for these things. So far, all the relevant projects formulated in Argentina have not been implemented. (*Physis* 1917: 81)

The only ray of hope named is the gift of Perito Moreno, whose national park idea was taken up and promoted by Bailey Willis. Willis was recruited as a U.S. specialist who celebrated great success in establishing Mount Rainier National Park and Glacier National Park in the USA. The Sociedad de Ciencias Naturales was familiar with the plans that Willis had submitted to the Ministry of Agriculture, as described here, and presented the project as "truly great" with the hope that this project would then be put into practice.

While *Physis* still took the fourth objective of the association, the promotion of national parks, very seriously in the first two volumes between 1913 and 1916 devoting long discussions to this topic, it is all the more astonishing that this discussion of national parks or other protected areas in the following volumes 3 to 5, which cover the period from 1916 to 1922, clearly declined until Lucién Hauman was to publish his groundbreaking essay "Para la protección de la naturaleza" here in 1922 (*Physis* 1916-1917; 1918-1919; 1921-1922).

The exception in the drought phase from 1914 to 1922 is a publication by Carl Curt Hosseus. Hosseus was from Hunsrück in Germany and had – similar to Willis – a contemporary, cosmopolitan imperial career as a biological discoverer. After his scientific and geographical studies in Freiburg, Kiel, Geneva, Leipzig and Berlin, he was closely integrated into the network of scientific associations in the German Empire. Among other things, he was assigned a botanical office in the German Colonial Service in Africa, but instead he ultimately conducted botanical excursions to Siam and Malaysia. He also worked in numerous European botanical academic institutions in Europe, including the renowned botanical garden of England, Kew Gardens. While there, he was invited by Dr. Huber to be his successor in the Museum Pará in Brazil, but Hosseus rejected this invitation. In 1912, however, he was invited to Argentina by the Argentine Ministro plenipotenciario Dr. Molina – through the mediation of the Argentine consul in Hildesheim, Prof. Dr. Hauthal. This invitation he accepted and in January he set out on the journey, only to never leave Argentina again – partly because of the turmoil of the First World War. In Argentina, Hosseus was commissioned in 1913 to make a first trip to explore the flora of northern Patagonia, and then in the summer of 1914 to make a second trip to explore the region around Nahuel Huapi. On both trips he was in close contact with Bailey Willis. For example, Hosseus was also paid 600 pesos by the Hydrological Commission for his second excursion in 1914 (Letter Emilio Frey to Bailey Willis, 30.9.1914, MP, Colección Frey, Biblioratо 7). And on the first trip he was accompanied in part by Carlos Gallardo, the governor of Rio Negro mentioned previously (Hosseus 2004: 6-13).

On the basis of his studies, Hosseus produced the first complete vegetation-geographical map of the Nahuel Huapi area (2004: 13), which was reproduced in his 1916 brochure. The German botanist arrived at similar results as the U.S. geologist Willis with regard to forestry use:

> In general, the wealth of these forests should not be exaggerated in the sense of practical use; without regeneration and destruction of the parasites, little wood will be harvested with the system practiced today. On the other hand, these forests, as natural beauty and promoters of rainfall, are of great importance for the whole area. Article 91 of the proposed Forest Act is to be welcomed with satisfaction: "Grazing is prohibited on the land of patagonic tax forests, where the low vegetation consists exclusively of young beech plants (Nothofagus sp.)." (Hosseus 1916: 33)

Hosseus' writing was a reference point for the botanical-geographical exploration and mapping of the region. In view of the new scientific debate on national parks in the journal *Physis*, the German botanist contributed the brochure "El proyectado Parque del Sud" in 1916. Although the vegetation-geographical main part of the brochure concentrates on Nahuel Huapi, Hosseus uses the introductory pages to position himself conceptually. In his booklet, Hosseus reports on efforts to establish national parks in five regions, namely

> the waterfalls of Iguazú, in Misiones; the lush virgin forests of Tucumán; a specific region in the province of Buenos Aires; the northern Patagoni-

> an region of the lakes in the border area of the Cordillera, especially Lake Nahuel Huapí; and Tierra del Fuego. (Hosseus 1916: 3)

And then immediately complained that despite numerous studies, "to this day, nothing has ever been put into practice" (1916: 3). Hosseus also criticized Willis' supposed geographical delimitation of the park around Nahuel Huapi. Here the frustration can be clearly heard through his writings of the national park projects in 1913 that ultimately politically failed. Legally Hosseus argues, "there is not even a law that would declare these regions OFFICIALLY national parks" (1916: 4, highlighting in the original). Only to then name this legal recognition as a national park as a political goal of his writing. In doing so, Hosseus – albeit without a discussion of national parks in other regions of the world – started to concretize the idea of the national park and to define two main goals:

> 1. To protect the region from unlawful intrusion by humans or animals and, as far as possible, from storms. 2. To open the protected region for visitors of the country's population and foreigners who come as tourists or mountaineers to see it. (Hosseus 1916: 5)

But even Hosseus, as a political activist for nature conservation, is ultimately unable to make his voice heard. His contribution lies above all in the scientific justification of the Parque Nacional del Sud project. In addition to the detailed collection of climate data – from precipitation, air pressure and temperature to relative humidity and winds – his focus is on the vegetation-geographical description, classification and mapping of the future national park (1916: 9-31). Hosseus then goes on to deal specifically with individual plants that are of particular importance for the region – also in terms of economic use (1916: 31-34). After hydrological mapping of the region as part of the Willis mission, Hosseus was now able to close the botanical-vegetation-geographical research gap (Hosseus 1916: 15).

In a way, the Hosseus brochure represents a dilemmatic culmination of previous efforts to establish a national park, especially at Lake Nahuel Huapi in northern Patagonia. This is where the scientific description of the region reaches its climax for the time being. But as important as Hosseus' work may have been scientifically, it had no consequences for the political implementation of national parks. While the work of Bailey Willis on the Nahuel Huapi National Park or the speech of Carlos Thays

– also sponsored by Francisco Moreno, among others – would push Bailey Willis' work to a broader public, this was not the case for the work of Hosseus. At the time of publication, the zenith of the national park debate had temporarily passed in 1913. The political framework conditions for the establishment of national parks were extremely poor due to the global political situation and the internal political disruption.

In view of the less than promising national constellation for the implementation of national parks, the Sociedad Argentina de Ciencias Naturales strengthened the regional and local dynamics for the promotion of nature conservation. Through its scientific networks, the Sociedad achieved a remarkably broad impact. It also succeeded in establishing contacts from the intellectual center in Buenos Aires to peripheral parts of the country. Thus, the first nationwide meeting of the Sociedad in 1917 did not take place in the capital, but in the northwestern provincial capital Tucumán. Here the politician and scientist Julio López Mañán, co-founder of the *Revista de Letras y Ciencias Sociales*, fought for the establishment of a national park in his region. On the occasion of the 100th anniversary of Argentina's independence, López Mañán proposed "the creation of an adequate 'reserve' in the provincial territory for the purpose of public welfare and recreation" (1971 [1916]: 117) in a wide area around the mountain San Javier and the Quebrada de Lules. The natural scientist Miguel Lillo had already made a similar suggestion in 1914. López Mañán justifies his initiative with the success of the national parks in the USA in preserving natural beauty, which would countered the progressive destruction of the environment in his own province. In doing so, he conceptually delimits the reserve from the urban parks or gardens:

> In contrast to parks and gardens, which are more or less urban and made for the enjoyment of the inhabitants of a city or a large estate, the reserve is intended for the nation. And, as its name implies, it tends to preserve everything that nature has created and can be counted among the best national sources of beauty, health or general well-being. (López Mañán 1971 [1916]: 118)

In contrast to the Wilderness discourse, he does not exclude human settlements in the reserve. Rather, he argues with Bailey Willis for the compatibility of nature conservation and human resource use.

In a region that partly coincides with the borders of the reserve proposed by López Mañán, Carl Curt Hosseus created the Aconquija National Park in 1928. Tucumán, with its citrus and sugar cane plantations, was considered the Garden of the Republic, and so the German botanist characterized the park in northwestern Argentina:

> [b]etween magnificent sugar cane plantations rise mighty forests, which house a rich vegetation of mosses, lichens, ferns, bromeliads, orchids and cacti. Orange trees and slender bamboo shrubs complete the attractive picture. (Hosseus 1928: 92-93)

However, it is not possible to reconstruct whether and when the park was established and in which particular boundaries. Presumably the protected area was first established at the political level of the province. It was not until 1995 that the Campos de los Alisos National Park was established in this area. In 2018, it was merged with other reserves to form the new Aconquija National Park.

The example of the Aconquija National Park highlights that, beyond the two emblematic park projects in the north and south of the country, there were a large number of regional initiatives to set up nature reserves. In this respect, it is too short-sighted to limit the initial phase of the national park movement in Argentina to the Iguazú Falls and Lake Nahuel Huapi alone. This can also be seen in the perception of contemporaries. In his emblematic article "La protección de la naturaleza en la república de Argentina" of 1922, Lucién Hauman adopts precisely this regionally differentiated perspective. Instead of limiting his view to the national showcase projects, Hauman broadens the perspective and, in keeping with the motto 'small is beautiful', also looks at smaller-scale approaches in other parts of the country. This included the "parque Aconquija" (1922: 290) established by the provincial government of Tucumán. In addition to the publicly established nature reserves, Hauman also positively highlighted private institutions such as the "reserva botánica" established by the brothers Rafael and Marcelino Herrera Vegas on the banks of the Río de la Plata (1922: 290-291). Accordingly, Martín Doello Jurado, also a founding member of the Sociedad Argentina de Ciencias Naturales, repeatedly advocated the establishment of national parks around Buenos Aires (1913) and in the La Plata region (1923). The Sociedad Argentina de Ciencias Naturales became active itself.

Thus Dr. Juan José Nágara, who won the Carlos Berg Prize of the Sociedad, donated part of the prize money to the Sociedad so that it could acquire land for the nature park on the outskirts of Buenos Aires (*Physis* 1915-1916: 184-185).

Although the establishment of national parks was certainly the main goal of the Sociedad Argentina de Ciencias Naturales, this society was at times able to establish itself as a point of reference in the academic debate and also in practical politics after the crisis of 1913 in the discussion of national parks in Argentina. After an enormous initial interest in the first issues of *Physis* and then a strong decline in the debate, a new interest in national parks emerged in the early 1920s.

The idea of how to establish parks had become scientific. And Hauman's proposals also showed a tendency towards professionalization, according to which natural scientists had to take over the establishment and administration of parks. The Belgian botanist proposed a three-stage procedure. Firstly, the parks should be based on scientific criteria, which necessitated an extensive scientific study, including inventory and survey by scientific institutions such as the Museo Nacional de Historia Natural or a botanical garden. In the second step, biologically valuable areas would have to be nationalized in the scientific interest. Thus, the common good would take precedence over private interests. And then – thirdly – the areas designated in this way would need to be expertly administered. For this task, the botanist preferred scientific societies with good financial resources. Alternatively, the municipalities could also be entrusted with this task if, as Hauman emphasized in the footnote, it is ensured that these work "under the exclusive guidance of a knowledgeable scientist". Hauman thereby implicitly brought into play the Sociedad Argentina de Ciencias Naturales itself as a possible candidate for the administration of national parks. Independent of this, however, he emphasized that national parks could only be managed by competent natural scientists. In the presentation of the Sociedad Argentina de Ciencias Naturales, the national park thus became a new occupational field for the small group of natural scientists.

1922: Park Foundation in the Tension Triangle of Natural Sciences, Radical Government and Aristocratic Elite

At the beginning of the 1920s there was an amazing revival of the national park idea at Lake Nahuel Huapi, which led to the official establishment of the Parque Nacional del Sud on 8 April 1922 by decree of President Hipólito Yrigoyen. The decree of the park's foundation states as its goal: "to prevent the destructive exploitation of the forestry empire" and at the same time "to preserve in certain regions the places of natural beauty with their original flora and fauna" (Ley 1922).

In this respect, this argument combines a Pinchot-like concept of sustainability of conservation with Muir-like wilderness protection in the sense of preservation. This decree, which is completely aligned with the US model, also addressed the protection of a large territory around Lake Nahuel Huapi. Specifically, an area of 785,000 hectares was placed under protection, while in 1919, the 7,500 hectares originally donated by Francisco Pascasio Moreno was increased tenfold to approximately 78,000 hectares, with the final park area increasing tenfold. The Parque del Sud, established in 1922, thus comprises more than 100 times the area originally donated by Moreno. In 1924 a further 43,000 hectares of land was integrated into the national park.

This makes the park almost as large as the Yellowstone National Park (8,983 km^2) and considerably larger than many other national parks in the USA. But this size – as Anasagasti et al. calculated in an essay on the National Park of the South from 1926 – must be put into perspective. For 85,000 hectares alone were allotted for the water areas of the lakes, of which 50,000 hectares were allotted to the Nahuel Huapi. In addition, there were 200,000 hectares within the park, which were privately owned – most of which (162,000 hectares) was in the hands of large landowners. Accordingly, the land area of the national park, which is state-owned, was still a respectable 500,000 hectares (Anasagasti 1926: 310).

The decree and enforcement of the national park around the Nahuel Huapi in 1922 was due to a quite new political-cultural coalition that successfully fought for the establishment of the national park, resulting in the formation of the civil society organization 'Comisión Pro Parque Nacional del Sur.' This honorary commission had a remarkable composition in that it included well-known representatives of Argentina's political and social elite. This included no less than the President of the Republic, Marcelo T. de Avilar, acting as Honorary President of the Commission. Honorio Pueyrredón, who had been Argentine ambassador to the USA since 1923 under Yrigoyen, Minister of Agriculture and Minister of Foreign Affairs, and under the follow-up government Alvear, was also represented as spokesman. It was Pueyrredón, in his capacity as Minister of Agriculture, who signed the decree establishing the National Park of the South. The Presidency of the Commission was held by Manuel Montes de Oca, former Minister of the Interior, and the great industrialist Carlos Tornquist acted as Treasurer. The individual members of the Commission belonged to the small porteñer elite. They knew each other personally and were also involved in other clubs such as the Jockey Club – the Argentine prototype of the British aristocratic club – or the recently founded automobile sports clubs such as the Automóvil Club Argentino, ACA, or the Touring Club Argentino.

In Argentina, this charitable social commitment was not unusual. For a successful political career, it was quite common at that time to become involved in supposedly non-political social clubs (Piglia 2014: 129; Losada 2007). Beyond political career planning, charitable work in clubs was regarded as a form of social distinction. As in the USA of the 'Gilded Age,' charitable social commitment belonged to the aristocratic tradition. But the USA was not considered a paradigm for the aristocratic lifestyle. The Argentine elite followed instead the European example of the English gentlemen. This was expressed in the founding of social clubs that indulged in sports such as golf, horse racing, polo, and automobile racing (Piglia 2014: 69, Losada 2007). The possibility of a state summer residence in a supposedly European setting – Nahuel Huapi as the Switzerland of Argentina is mentioned specifically – certainly also played a central role in a geopolitical imaginary focused on virility, adventurousness and the desire to conquer in order to fuel the commitment of these men, with Patagonia being the last colony to be conquered.

What was special about the Pro-Parque Commission, however, was that it built the bridge between the aristocratic elite in Buenos Aires and the local emergent elite at Lake Nahuel Huapi. For many of the aristocratic members and friends of the commission, the barely developed northern Patagonia represented a prominent field of activity. Anasagasti and the co-founders of the Aero Club, Aarón de Anchorena and the famous pilot Jorge Newbery, had settled in Patagonia or established summer residences there. Aarón de Anchorena's love for Patagonia went back a long way. In 1902 he travelled Patagonia with the support of the then President Julio Argentino Roca, where he met, among others, Francisco Pascasio Moreno and he was particularly impressed by the island of Victoria in the middle of Lake Nahuel Huapi, where he was granted access for life from 1907, but gave up this right of use in 1911. Ángel Gallardo – later president of the commission – had also acquired land for the construction of a summer residence at Nahuel Huapi (Frey to Gallardo, 1.12.1928; MP, Colección Frey; Biedma 2003: 192). The same applies to the Scottish immigrant Ernesto Jewell, who at that time was also president of the aristocratic Hurlingham Club, which had been founded in 1888 in Argentina after his famous English model. Likewise, Anasagasti's nephews, Luis and Fermín Ortiz Basualdo, both recognized personalities of the aristocracy in Buenos Aires, also acquired possessions on the banks of the Nahuel Huapi. They built their summer residence in the Huemul region, where Argentine President Mauricio Macri spent his holidays in the summer of 2018. Obviously, the aristocratic charm of the region has not yet faded.

One of the central transregional points of convergence for the national park idea was the industrialist Horacio Anasagasti, who also acted as Secretary General of the Commission. Anasagasti was one of the first car manufacturers and racing drivers in Argentina to co-found the Automóvil Club Argentino (1904), the Touring Club Argentino (1907) and the Aereo Club Argentino (1909). He also established a channel of communication into the academic field on the basis of his vice-presidency of the Sociedad Científica Argentina (1909-1911).

This assessment of the importance of these aristocrats with good connections in northern Patagonia is also shared by the central figures of civil society in Bariloche, such as Emilio Frey. In addition to Aarón de Anchorena and the big landowner Jewell, Frey considered Fernández Beschtedt to be a central person who supported the founding of the park and also exerted influence on Yrigoyen. This influence led so far that in the founding decree of the Parque Nacional del Sud ad nombran Emilio Frey was appointed as national park representative (Letter Frey to Fernández Beschtedt, 31.8.1928, MP, Colección Frey, Caja 8).

On 14 April 1924, two years after the formal establishment of the National Park, the Pro-Parque Commission, initially a civil society commission, was officially commissioned by President Alvear, by decree, to draw up a general plan for the management and regulation of the National Park. This decree reflected the ever-closer connection between the elitist, charitable Pro-Parque Commission and the state institutions. In 1926 Horacio Anasagasti, in his function as Secretary General of the Comisión honorifica "Pro Parque Nacional del Sur," prepared the report *El Parque Nacional del Sur*. What is remarkable about this report is that Anasagasti succeeded in combining the more recent political efforts of the aristocracy with the scientific efforts of previous years. The report contains a geological chapter by Anselmo Windhausen, a German-born geologist, and a vegetation-geographical chapter by Carl Curt Hosseus. In this chapter, Hosseus emphasizes the unique combination of botanical diversity and scenic beauty and, recalling his own excursions in 1913 and 1914, provides a declaration of love to the National Park of the South:

> There are few places in the world where a relatively small area such as this park can boast such a great variety of plants, and it is rare for a botanist to find a study area as beautiful as Lake Nahuel Huapí. (Hosseus 1926: 287)

However, the report not only bridged the academic gap, Emilio Frey himself also contributed. Although he was a surveyor and a member of the expedition under Francisco P. Moreno and Bailey Willis, he had unique local knowledge, although neither a scientist nor an aristocrat. The participation in the Anasagasti report can therefore only be understood as an indication of the central role Frey played as the first director of the park and in the establishment and implementation of the National Park of the South, which should not be underestimated.

Around the year 1922, however, the report *El Parque Nacional del Sud* was not the only document to put the debate on national parks back on the political agenda. In this phase of the political revival of the national park and nature conservation initiatives in Argentina, Lucién Hauman's "Para la proteccion de la naturaleza," was a programmatic essay on nature conservation. Published in 1922 in the magazine of the Sociedad Argentina de Ciencias Naturales *Physis*, Hauman developed a differentiated understanding of nature conservation. While the group around Benito Carrasco – coming from the Thays area – mainly focused on the protection of natural monuments, especially from the perspective of heritage, Hauman criticized this approach writing, "not only famous attractions or curiosities that are more or less legendary can be endangered, but also the most frequent and thus most interesting aspects in which we live" (1922: 284). And more precisely he adds, "and you can't just believe that only special or classic places deserve protection" (1922: 290). Hauman is also cautiously critical of capitalism, although these criticisms would not reach the moral sharpness of the public contributions of the Sociedad Forestal in the 1910s:

> Everyone knows that the progress of so-called civilization – or rather, the maximum exploitation of natural wealth – profoundly transforms the environment: Cities expand into fields, forests retreat, wetlands dry out, rocks become quarries, plains are covered with crops, and wherever there is vegetation, herds graze to saturation. (Hauman 1922: 283)

But in order to then relativize his critique again in the sense of universal progress, he adds, "And so it must be" (Hauman 1922: 283). While in the USA and also in the early writings of Eugène Autran the idea of wilderness still resonates, Hauman already disillusioned proceeds from the "irreparable destruction of the primitive aspect of nature" (1922: 583-584).

It was precisely from this that a moral imperative for nature conservation was derived that emphasized the intrinsic value of nature. Hauman's undertone is also critical of consumption and commodification when he writes, "Protection must be extended to that which is threatened with disappearance, even if it would never deserve the attention of tourism authorities" (Hauman 1922: 290). Hauman defines a normative imperative for nature conservation that is also justified as a duty to protect species for future generations. In contrast to utilitarian sustainability concepts, Hauman does not focus on sustainable use, but on conservation – contrary to the tendency of species extinction (1922: 285). For Hauman, ecological degradation or even ecocide is "*complete and final*" (emphasis in the original). For Hauman, it results from the inevitable process of civilization, which is entirely dedicated to the idea of progress. But in a dialectic of civilization, the unavoidable consequences of modern society, such as the threat of ecocide here, result in the task of civilization, indeed the moral duty of civilization, to protect nature. Accordingly, the Belgian scientist formulates:

> It is our duty to preserve for future generations the historical memories of the primitive landscape, to create large or small strategically scattered "reserves" in which the primitive associations of plants and animals, the first inhabitants of the land, can be preserved under natural conditions. (Hauman 1922: 285)

This meaning of intergenerational responsibility is certainly borrowed from the ideas of sustainable forestry in Clausewitz, but here Hauman has expanded the concept in the sense of an ecological ethic that is quite close to the concept of sustainability as it was propagated at the Rio Conference in 1992.

Hauman systematically justifies the necessity of establishing nature reserves from four perspectives: the scientific point of view, the artistic, the sentimental, and the utilitarian. From the first point of view, he complains that the objects of investigation – the biota – of the natural scientists are being destroyed. They must therefore be preserved in the service of science. Secondly, he sees nature as a quasi-primordial creative force that is also passed on to the younger generations for their education. Therefore, it is central to enable the following generations to experience nature in order to produce new artists – in a Beuysian understanding of art in which every human being is an artist (1922: 285-286). Third, he criticizes simplified forms of patriotism that are based on the separation of nature and history and thus neglect an important part of experience, namely "the love for the parts of the world that surround us, for the stones, for the plants, for the animals that form them". But these – according to Hauman – are also an essential "form of love for the fatherland" (1922: 287). Fourthly – and consciously lastly – the biologist then also mentions utilitarian aspects. Here he cites as an example a number of economic consequences that even the smallest living beings can have

> tiny mushrooms, to which a banker would not give a second thought, cost their land more money every year than it will ever use in its business during its lifetime. (Hauman 1922: 287)

In his search for best-practice samples for nature conservation, Hauman rejects the U.S. model, which has been very popular and well-known in

Argentina to date. He greatly appreciates the U.S. model, but considers it too demanding for the Argentine context and therefore not imitable. Indeed, he warns,

> [i]t is necessary to say and to repeat again and again that one should not only think of grandiose works, of huge national parks whose admirable projects attract attention, but which do not go beyond the project status. (Hauman 1922: 290)

Instead, for Argentina he has "a more modest and perhaps more effective job" in mind, as it has already been implemented in other countries. Hauman uses the publication of his Belgian compatriot Jean Massart *Pour la protection de la nature en Belgique* from 1912 as a guideline (1922: 288). In all these cases, it is the modern man, or better the expert scientist, who should organize the protection of the environment in the sense of the idea of stewardship.

Based on this discussion of worldwide efforts to establish nature reserves, Hauman for the first time draws a balance of Argentina's efforts under the chapter heading "Lo que se hizo en Argentina," which, however, is more of an inventory. Although Francisco Moreno's donation is first listed as the founding act of the national parks, the debates surrounding the establishment of the national park around the Iguazú Falls are almost completely absent here.

The distinction made by Lucién Hauman between a natural monument and a national park was also taken up by Carl Curt Hosseus in a 1926 petition to the Universidad de la Plata to establish a national park for the center. Here, the botanist of German origin distinguishes between local protection and the protection of large areas. He assigned the first approach to the protection of monuments, in which "isolated or interconnected objects" legally protected the biological, geological or historical-archaeological importance or particularly endangered animal species. According to Hosseus, the second case, which concerns the national parks is more complex because of its size. An explicit example of this form of nature conservation were the 13 national parks established in the USA to date (Hosseus 1926: 558). With regard to the National Park of the South, Hosseus sees this criterion fulfilled, since the 1922 boundaries drawn "correspond to the natural delimitation of a botanical entity that encompasses all important phytogeographic elements within these latitudes" (Hosseus 1926: 286).

Horacio Anasagasti argues differently. The secretary-general of the Pro-Parque Commission came up with an interesting combination of both approaches for the National Park of the South. He based his approach on the report that gives "the example of the great state of the north of our continent," which "must show us the way in the essential aspects of this project" (Anasagasti 1926: 302-303). The USA – according to Anasagasti – makes the technical distinction between "national parks" and "national monuments," whereby the latter may well be located in the parks. With regard to the parks, Anasagasti sets himself apart from the concept of the natural monument, which was particularly influenced by Conwentz in Germany, and gives priority to the U.S.-American idea of combining nature conservation, tourism and moral-physical edification through nature. The notion that the experience of nature calms the nerves and strengthens the organism (1926: 302) is clearly reminiscent of the Roosevelt-Darwinian idea of "strenuous life" (Brinkley 2010: 368-395).

For the Park of the South, this means that it should be developed for tourism and connected to the transport network – following the arguments of Bailey Willis. A limited agricultural use should also be possible, as should tourist and urban construction projects. In order to maintain and preserve the "natural and untouched character of nature" (Anasagasti 1926: 311), reserves should be designated inside the national park where humans would have no access. Specifically, Anasagasti proposed two such reserves. In addition, the park was to be divided administratively into different districts, each of which would be under the control of a park ranger. This subdivision corresponded to the zone model that is also common in today's national parks in order to define different intensities of use and protection.

The implementation of the decree establishing the National Park of the South initially progressed rapidly. Emilio Frey, engineer and geographer of the General Direction of Lands, was commissioned by the decree to draw up a provisional regulation for the national park (1922). At the same time, he worked out the surveying aspects of the demarcation of the National Park, which was now considerably enlarged compared to the original donation (Polar, Esteban, June 27, 1964, MP, Colección

Frey, Caja 4). On December 1, 1922, Frey presented a handwritten regulation for the National Park in his function as "Commissioner of the Parque Nacional del Sud". It prohibited the felling of wood in the state lands, the killing of wild animals, and it regulated slash-and-burn and defined emergency measures in case of forest fires. To this end, the park was divided into seven zones, each of which was assigned responsible park rangers (Frey, Reglamentación provisoria, 1.12.1922, MP, Colección Frey, Bibliorato 8).

The establishment of the first national park in Argentina also gave the country wings on the international stage. In 1924, on the occasion of the Panamerican Congress in Lima, the Argentinean delegate, the General Director of the Ministry of Agriculture, José León Suárez, emphatically called on the assembly of American states to establish nature reserves.

Despite this astonishing dynamic for the establishment of the national park at Nahuel Huapi around 1922, the process of further administrative implementation came to a standstill at the end of the 1920s. The papers of the director Emilio Frey show that he did not carry out the work of the park administration in formally clarified employment relationships and institutional ties, but rather on the basis of personal conviction in a legally-administrative grey zone. In a 1928 letter to Dr. Ángel Gallardo, director of the Museum for Natural Science and later president of the Comisión de Parques Nacionales, with whom Frey had a confidential and friendly relationship, he wrote in regard to the implementation of the decree of 1922 and the provisional regulations of the same year:

> But we have made no progress, everything has stopped at the decree. I am still dependent on the Directorate-General for Land, in my capacity as Territorial Commissioner for the Rio Negro territory, with headquarters in Bariloche, which has allowed me to take care of the Park's interests and supervision. (Letter Frey to Gallardo, 1.12.1928, MP, Colección Frey, Bibliorato 8)

In 1928, in the face of the second presidency of Hipólito Yrigoyen, Emilio Frey was committed to approaching this president, who had the merit of establishing the National Park during his first presidency. In concrete terms, he called for the institutionalization of the park with "its internal statute, its regulations and its budget, so that it has authority, autonomy and government" (Letter Frey to Montes de Oca, 12.12.1928, MP, Colección Frey, Bibliorato 8).

A suitable person to take over the management of the National Park was also to be considered. Commissioner Ernesto Jewell, who lived on the bank of the Nahuel Huapi, suggested – probably influenced by Frey – Fernández Beschtedt, a confidant of Yrigoyen, while the latter had Jewell in mind. In view of this debate, Frey refers to the strategic importance of the director for the institutionalization of the national park when he states, "I believe that as a director we will need a good puppet in front of the new government, connected to its men, to do something and free the park from its nostalgia" (Letter Frey to Fernández Beschtedt, 31.8.1928; MP, Colección Frey, Bibliorato 8). Frey himself had Yrigoyen's trusted friend in mind, and so he asked him directly "Wouldn't you be the right person for the job yourself?" In order to make Fernández Beschtedt's post palatable, Frey presented him with a far-reaching vision for the administrative reorientation of the national parks. In view of the establishment of the Parque Nacional del Norte and another in Tierra del Fuego, Frey argues that: "there would have to be only one commission to take care of the management of all the parks". In specific terms, Frey presents the following:

> Under this commission would be the section offices of the national parks and a permanent office in Buenos Aires with director, secretary, cashier and other necessary personnel. Each National Park will in turn have a delegate or director, staff and budget. (Letter Frey to Fernández Beschtedt, 31.8.1928, MP, Colección Frey, Bibliorato 8)

In addition to this administrative modernization, Frey also aims to further professionalize the national parks:

> For each park there must be a specialist who has worked in European or North American parks; the rangers must be trained persons who have knowledge of the species and the maintenance of the forests. (Letter from Frey to Fernández Beschtedt, 31.8.1928, MP, Colección Frey, Bibliorato 8)

The subsequent discussion process around these regulations and their enforcement cannot be deduced from the sources; the earliest printed public announcement is dated February 1, 1928 (Reglamentación Provisoria, 1.2.1928, MP, Colección Frey, Bibliorato 8).

There is no doubt that Emilio Frey, whose moral values Bailey Willis once described as being "intensely loyal to Moreno's ideals" (Willis 1947: 48), was doing everything in his power to realize Francisco Pascasio Moreno's vision despite the adverse circumstances. In the face of Moreno's lack of instructions, Frey increasingly pushed ahead with a national park project that, although in the tradition of Moreno and Willis, also received a very clear personal signature from Emilio Frey. In particular, Frey drafted the basic organizational ideas of the National Park administration, which were then implemented from 1934 onwards with the National Park Act. Despite the lack of institutionalization at the national level, the National Park was still in the process of being implemented at the local level. A minimum number of staff were also available to monitor the park rules. Frey worked in the national park with six assistants, for whom the Yrigoyen government even had four park ranger huts built in 1928 (Letter Frey to Montes de Oca, 12.12. 1928, MP, Colección Frey, Bibliorato 8). Social and tourist development was also progressing with the construction of a school and the expansion of the hotel infrastructure.

Thus, the establishment of the park stimulated new economic activities in the emerging tourism sector in Bariloche. In 1930/31 the German-born inhabitants of Bariloche, Hans Hildebrandt and Otto Meiling, published a tourist guide in which they advertised the region around Nahuel Huapi as an idyllic place in which the city dweller, who was rushed by nervous city life, could relax. Hildebrandt had founded a tourism office in Bariloche at that time and Meiling worked as a mountain guide and later also as a ski instructor. Furthermore a hotel called Parque Nacional was founded here and in June 1927 also a brewery "Parque Nacional". But the operators, Esteban Zufiaur and Juan Antonio Leberle, had to stop production as early as 1931 – probably due to increasing competition from the railway connection.

"Nervous Buenos Aires vs. Peaceful Nahuel Huapi"
(Hildebrandt and Meiling 1930/1931).

As in other countries, the railway company – here the Ferrocarriles del Estado (FF.CC.) – was one of the leading forces in the promotion of tourism. In 1928 they published the elaborately designed brochure *Parque Nacional del Sud* to promote tourist destinations in the National Park. The brochure, richly illustrated with photographs and artistic illustrations, also contained practical travel information such as maps and timetables, as well as advertisements of hotels, tourism providers and other service providers of interest to tourists. Horacio Anasagasti has also enriched the magazine with a short article on the road network for automobiles (FF.CC. 1928).[5] In the following year, the brochure *Parque Nacional del Sud* (FF.CC. 1929) was completely overhauled and redesigned.

5 For the cover illustration of the 1929 FF.CC. brochure see the front-cover of this book.

Cover, FF.CC., “National Park of the South”, 1928.

The national park idea took place not only in the south and north, but from north to south. In this sense, the successful establishment of the Parque del Sud in 1922 also brought about the revival of further regional initiatives. Carl Curt Hosseus, for example, pushed the founding of a Parque Nacional Central in his hometown of Córdoba (Hosseus 2004: 21). Hosseus convinced the university congress of the University of La Plata to “recommend the enactment of a law or decree by the nation’s executive to create the central national park in the mountains of Córdoba” (1926: 558). This step is of particular interest for the genealogy of the national parks in Argentina, since traditional historiography has concentrated solely on the emblematic national park projects in the north and south. However, this petition makes it clear that nature conservation efforts were far more complex and regionally differentiated. In addition to the parks in the north and south, there was also to be a park in the center.

The geopolitical imagination of being in the center of the country is an essential part of the justification for the establishment of a park in the Sierra de Córdoba: "I must point out that a central region of the country with its special characteristics is not yet envisaged as a central national park" (1926: 558). The centrally located region of the Sierra de Córdoba is also referred to as the "heart of the country" because of its infrastructural connections. In addition, of course, its special natural environment is also referred to as the region that "is home to the most characteristic natural monuments. Its climatic conditions, minerals, flora, fauna and beauties are very diverse: Hills, forests and waterfalls" (1926: 558).

Hosseus is also responsible for the international promotion of the Central National Park. In 1928 he published the results of his work in the German journal of the association *Naturpark e.V.* in the report "A new national park in Argentina," which is illustrated with two photos of the Argentine Ministry of Agriculture – one of the Central National Park and one of the National Park of the South. Hosseus praises the park on 150,000 ha at the Sierra de Córdoba in highest tones where he writes, "A special adornment of the park forms the 2850 m high Cerro Champaqui with its characteristic highland sea". At the same time he – quite atypical for the nature conservation discourse in Argentina, which is otherwise related to landscape or forest protection – emphasizes the protection of fauna, "At its heights the condor and the guanaco still live today; there is the danger that they will be completely exterminated" (Hosseus 1928: 93).

As far as the regional dynamics and diversity of nature conservation efforts are concerned, the petition is also highly interesting in a second respect. For Hosseus begins – hardly surprisingly from today's point of view – with: "There are already two national parks in the Republic of Argentina". But there is a surprise when naming them. He omits the Iguazú Park and names first a national park in Tucumán – the Aconquija Park (Hosseus 1928: 92), then – not surprisingly – Nahuel Huapi. Hosseus thus follows the 1923 argumentation of his colleague, Sociedad Argentina de Ciencias Naturales Lucién Hauman. The immediate surroundings of the Iguazú Falls, despite the early attempts to protect them, are only listed as an area reserved for the creation of a national park. In addition, Hosseus points out, as Hauman did before, the lack of enforcement of the protection efforts at the Iguazú Falls. In the discussion in the University Council following Hosseus' lecture, engineer Juan B. Marchionatto then made it clear that – in addition to the cases mentioned – there had also been efforts to set up nature reserves in other places such as the Delta des Paraná or the island of Martín García (1926: 558). Another indication of the local diversity of nature conservation efforts in Argentina in the 1920s.

1934: Exequiel Bustillo and the Institutionalization of National Parks

The decisive step towards the final institutionalization of the national parks in Argentina took place under the authoritarian government of General Agustín Pedro Justo in the 1930s. In 1930, the liberal president Hipólito Yrigoyen was overthrown by a military coup. The military under the leadership of General José Félix Uriburu took power and represented the interests of the old oligarchy. As Uriburu increasingly turned to Italian fascism, he was no longer tenable for the conservative elites and was replaced in the 1932 elections – also on the basis of electoral fraud – by the pro elite militarist and politician Agustín Pedro Justo, who had supported the coup at the time. This epoch of the 1930s was also called *década infame* because of the democratically illegitimate governments. As a result, the traditional oligarchy in Argentina was once again firmly at the helm from 1930 to 1938, steering the country's fortunes.

This phase of conservative rollback was helpful for the implementation of the national parks in so far as many political representatives of the traditional oligarchy were committed to the national park idea. Nevertheless, there was also a shift in the national park discourse during this period. This applies especially to the spokespersons of the debate. The naturalistas, who were committed to a scientific justification of the protected areas and – like Lucién Hauman – had spoken out in favor of the political-administrative implementation of nature reserves by experts from the natural sciences, lost importance during this time and the political lobbyists from the oligarchy gained the upper hand. Although natural scientists and representatives of the elite were closely linked both ideologically and personally since the beginning of the national park initiatives, the elite's claim to leadership had now became apparent.

This shift is all the more striking in view of the fact that, precisely at the time of the political upheaval in Argentina, there were also momentous personnel crises in the Pro-Parque Sud Commission. The chairman of the commission, Horacio Anasagasti, suffered a fatal heart attack on 8 April 1932 – curiously enough on the 10th anniversary of the park's foundation. The former Minister of Agriculture, Antonio de Tomasso, took over his post, but he passed away in 1933. His successor, Manuel Augusto de Oca, died in January 1934, so his term of office was also short. The same applies to Ángel Gallardo, the respected scientist and politician who also headed the Commission de Parques Nacionales for only a few months until his death on 13 May 1934.

In his function as Minister of Agriculture, de Tomasso reorganized the Commission in 1933. The Commission was originally set up in 1924 to deal solely with the establishment of the Park of the South, but was now restructured into a nationally operating Comisión de Parques Nacionales (Bustillo 1997: 89). At the time, this meant in particular that this commission now also had the responsibility to establish the Iguazú National Park (Freitas 2018: 114). After Gallardo had taken over the presidency of the commission, a significant power vacuum arose after his death. Surprisingly, the young lawyer and representative of the traditional oligarchy, Exequiel Bustillo, assumed the office of president. Unlike Ángel Gallardo, Carl Curt Hosseus, Lucién Hauman or Carlos Thays in Argentina, Federico Albert in Chile or John Muir and Giffort Pinchot in the USA, Exequiel Bustillo was not a natural scientist. In fact, he was not even scientifically educated. Bustillo remarked self-critically that his botanical ambitions were quite limited:

> I lacked passion and scientific background in the field of botany, so I could only distinguish a eucalyptus from an acacia or a casuarina... And yet, from that moment on, the most beautiful forests of the Republic remained under my supervision. (Bustillo 1997: 97)

Especially through his autobiographical writing *El despertar de Bariloche* (first edition from 1968), which Bustillo had written on the basis of his private archive, and which is now in the Archivo General de la Nación, Exequiel Bustillo successfully made a name for himself as the actual creator of the national parks. Bustillo, who was deeply rooted in the aristocratic oligarchy of Buenos Aires, presented the earlier park initiatives of Francisco Pascasio Moreno, Carlos Thays and Bailey Willis – while concealing the works of Carl Curt Hosseus and Lucién Hauman. With regard to the Nahuel Huapi National Park, Bustillo – consciously or unconsciously – ignores the work already done on site in the National Park of the South, which had existed since 1922. Bustillo certainly referred to the outstanding qualities of his collaborator Emilio Frey, whom he describes as “the greatest luxury” especially because of his merits on the expeditions of Moreno and Willis (1997: 101). But despite this eulogy on Frey, Bustillo conceals that Frey was not only an official of the

Dirección de Tierras at that time, but in this function the official director of the National Park of the South, established in 1922 and that in this capacity he had carried out various works to set up the park and improve its infrastructure. Recent works – especially regional historical research from Patagonia – have rightly pointed out that the Bustillo myth of the park's foundation must be deconstructed and regionally-historically corrected (Bessera 2008, Navarro and Vejsberg 2009, Navarro 2010).

The great merit of Bustillo, however, lies in the enforcement of the national park law, which had been long demanded. As Bustillo vividly describes in his autobiography, the passing of the law in parliament succeeded primarily on the basis of Bustillo's personal networks in the conservative oligarchy. The law itself was then drafted under the direction of Ángel Gallardo and later Exequiel Bustillo, Gustavo Eppens and Antonio Lynch (Bustillo 1997: 91). Below this management level, the legislative decree itself seems to have been formulated with the collaboration of Rodolfo del Castillo, who was in close contact with Emilio Frey. Thus, del Castillo wrote in September 1933 to Frey, "I am glad my project was not changed by a single comma. (meaning the draft for the decree of law, O.K.)" so that the commission could then pass it on to Minister Duhau. There was support from the very top:

> The President also supports the initiatives: At the special hearing agreed by the Commission, he spoke with full knowledge of Nahuel Huapi and found the solutions proposed by the Committee very appropriate. (Letter Rodolfo del Castillo to Emilio Frey, 8.9.1933, MP, Colección Frey, Bibliorato 8)

Frey was already a candidate for the superintendency of Nahuel Huapi National Park. At the end of September, Gallardo, then chairman of the commission, and Bustillo intensified their political pressure to enforce the law. This draft then only contained two minor modifications (Letter Rodolfo del Castillo to Emilio Frey, 29.9.1933, MP, Colección Frey, Bibliorato 8).

On 30 September 1934, law number 12103 – the Argentine National Park Law – was enacted. With this law, the Nahuel Huapi National Park and the Iguazú National Park were regulated and/or established. In addition, a national park authority, the Dirección Nacional de Parques Nacionales (DPN, National Direction of National Parks), was established, which was to be responsible for all national parks, including future ones. Specifically, it was equipped with the power of disposal over public lands in national parks and reserves. For the implementation of this power of disposal, General Justo issued Decree 55177 in which the Dirección de Tierras of the Ministry of Agriculture was instructed to hand over all Land concessions and forestry areas within the national parks to the DPN by end of November 1935. Similarly, all the facilities of the Ministry of Agriculture within the National Parks were transferred to the DPN, with the exception of the Dirección de Tierras building in Bariloche and the fish farm on the banks of the Nahuel Huapi. The land rights of private individuals within the parks initially were to remain protected, but DPN could ask the police to evacuate illegal intruders (Decree 55177, AGN, Legajo Bustillo, Caja 3344).

In the Iguazú National Park, in the northern province of Misiones, the transfer of the military area to the new national park authority was particularly problematic. The main point of contention was the Ayarragay Hacienda, which the military had taken possession of in 1928, and which they had been managing ever since as Campo Nacional de Iguazú. In August 1935, the president issued a decree after the military was to keep 20,000 of the total 75,000 hectares of latifundium, while 55,000 hectares were transferred to the DPN. But even this area was controversial, since parts of the military wanted to use 50,000 hectares of this area for a mega project of agricultural colonization, in the course of which 2,000 Argentine families were to be settled in the region. It was not until 1941 that the territorial disputes with the military were settled and resolved by presidential decree so that 5,000 hectares were released for agricultural colonization, while the rest of the land remained in public hands (Freitas 2018: 117). This defined the borders of the two emblematic Argentine national parks in the north and south of the country.

Exequiel Bustillo's geopolitical vision, however, focused on Patagonia. His aim was to complete the project of colonizing this southern region of Argentina, which had begun at the end of the 19th century, and to secure the border line with Chile's neighbor established in 1881. In this sense, following the institutional safeguarding of the Nahuel Huapi National Park, the newly established Dirección de Parques Nacionales (DPN) focused on other regions of Patagonia. Accordingly, the tourism development and colonization of the southern regions of the country by national parks was intensively promoted as early as 1936.

Glaciar Perito Moreno. DPN 1937. *Nuevos Parques Nacionales*. Buenos Aires.

In the brochure *Nuevos Parques Nacionales*, lavishly designed with picturesque photographs, the DPN presents the results of several expert commissions that had investigated the establishment of reserves for the purpose of establishing national parks in the national territories of Neuquén, Chubut and Santa Cruz as early as 1937 (DPN 1937a). DPN technicians presented detailed reports including detailed maps and geographical descriptions of the future national parks Lanín, Los Alerces, Perito Francisco P. Moreno and Los Glaciares – in the deep south of the country.

In 1943, the Section for Conservation of the Panamerican Union's Office for Agricultural Cooperation asked DPN for a list of all Argentine national parks (Letter Panamerican Union to Bustillo, 23.9.1943; AGN-Legajo Bustillo, 3348). With the elaboration of the Panamerican Convention for Nature Conservation in 1940, the Panamerican Union had already distinguished itself as a central authority for the dissemination of nature conservation in the western hemisphere. The information requested here should now lead to a complete inter-American list of national parks. The Pan-American Union already has the following parks established in Argentina: Nahuel Huapi, Lanín, Los Alerces, Perito Francisco P. Moreno, Los Glaciares, Iguazú and Angel Gallardo as well as the Reserva Los Copahues. Only detailed information is missing, like the year of the establishment, the size, control organization, budget, as well as special flora or fauna.

In his reply to William Vogt, head of the Panamerican Union's nature conservation department, Bustillo provides precise information. He makes it clear that only the National Congress can decide on the establishment of national parks. In October 1943, Argentina had only two national parks, Iguazú and Nahuel Huapi, established in 1934. On May 11, 1937, the executive branch issued a decree that established reserves on state territories as national parks in the founding process and transferred the administration of these territories to the DPN. A corresponding law on the establishment of the parks Lanín, Los Alerces, Perito Francisco P. Moreno and Los Glaciares had been passed, but had not yet been adopted. According to Bustillo, the current political upheaval in Argentina made it impossible to adopt the bill because the Congress had been dissolved. And the Ángel Gallardo National Park did not go beyond the status of a project idea (Letter Bustillo to Vogt, 15.10. 1943, AGN, Legajo Bustillo 3348).

Finally, the four reserves intended for the establishment of the National Park were officially established as National Parks on 29 May 1945, together with the Laguna Blanca National Park in Neuquén, under the dictatorial regime established in 1943.

For the organization of the parks, further regulations and planning were necessary, which dealt with the regional and ecological peculiarities. The internal regulations of the Nahuel Huapi National Park, which were drawn up in parallel with the National Park Act, were eagerly commented on in December 1933. It is interesting to note that, contrary to the authoritarian spirit of the time, Frey wanted to curtail the decision-making power of the Superintendent, a position for which he himself was intended (Letter from Emilio Frey to Rodolfo del Castillo, 20.12. 1933, MP, Colección Frey, Bibliorato 8). At the same time, new work was begun on Nahuel Huapi Park. Park rangers were hired, and a car was bought for the local park direction.

With the new boom in the institutionalization and formalization of national parks at the national level, a renewed debate about the importance and function of national parks also began. One of the primary goals of this debate was to define what a national park is. In contrast to the debates from the beginning of the 20th century – and above all the 1920s – the spokesmen were no longer scientists, but politicians of the Argentine oligarchy. Exequiel Bustillo writes self-critically about his expertise in the field of nature conservation policy:

> In its technical aspects I did not know much about the subject; but I had an idea of the regional needs, and above all I was convinced of the urgency of nationalizing and occupying a border over which we exercised purely nominal sovereignty. (Bustillo 1997: 91)

Instead, he refers to his legal expertise writing, "My main sources were the national park laws of the USA and Canada. Also the law for national territories, because at that time our only two parks were within its geographic area" (Bustillo 1997: 91-92). In keeping with the conservative *Zeitgeist* and the increasingly popular fascist tendencies in Argentina, Bustillo is a striking accentuation of the geopolitical function of the national park. Accordingly, he defines a national park in his autobiography as being

> an official body which, in addition to its specific function of preserving natural beauty and making it available to the people, also pursues the ambitious and difficult goal of strengthening our territorial sovereignty in its region. (Bustillo 1997: 10)

The almost complete absence of environmental protection in the discourse of the National Park Authority is a striking feature. In March 1942, the Argentine newspaper *Ahora* conducted a written interview with Exequiel Bustillo on the significance of the national parks. Neither in the questions of the journalists, nor in the answers of the director of the national park authority do nature conservation aspects occur. Instead, Bustillo defines the parks as a development engine that has only just begun:

> The national parks are still no more than an embryo. There are still more and better ways missing. More hotels. More landing stages. More hospitals and first aid rooms. More schools. With a more organic plan and greater resources, an extensive reforestation programme needs to be tackled. This is the foundation, as in the future it will be one of the most important underpinnings of the local economy. (Letter Ahora to Bustillo, 3.3.1942; Letter from Bustillo to Ahora, no date, AGN, Legajo Bustillo, Caja 3347)

Nature conservation, understood as the aesthetic preservation of the landscape, is primarily seen as functional for tourism. Bustillo, in his speech on January 20, 1937, at the opening of the road network in the Nahuel Huapi National Park, emphasized that the goal of nature conservation was to put the landscapes into value "so that they may serve as an attraction for the great tourist crowds". With regard to patriomonio, this discourse is based on the natural heritage debate in the USA. In the absence of relevant historical buildings in the field of cultural heritage, the natural landscapes are defined as shaping the national heritage. Alfred Runte analyzes this in his classic study of the national park idea in the USA published in 1949:

> Unlike established, European countries, which traced their origins far back into antiquity, the United States lacked a long artistic and literary heritage. The absence of reminders of the human past, including castles, ancient ruins, and cathedrals on the landscape, further alienated American intellectuals from a cultural identity. (Runte 1979: 11)

With this in mind, Bustillo argues that Argentina should also pursue this strategy when he writes, "It is the same thing that new countries like ours must follow, which lack the attraction of the old European civilizations," where Bustillo, for example, has in mind the 400 castles along the Loire valleys, a region that was declared a World Heritage Site by UNESCO in 2000 (Bustillo, Speech of 20.1.1937, AGN, Legajo Bustillo, Caja 3354).

And despite this emphasis on monumentality, wilderness also becomes a compelling argument for the establishment of national parks. Thus, Bustillo rejects the plans to establish a park close to the city in Buenos Aires, since it does not meet the wilderness criterion. Bustillo writes accordingly to the initiator of the park, Roberto Barili:

> We deeply regret that the establishment of the park that you are planning near the lagoon of 'Los Padres' near this beautiful seaside resort is not in accordance with the law that applies to us, as it is an artificial park and not a national park. (Letter Bustillo to Roberto Barili, 17.9.1941, AGN, Legajo Bustillo, Caja 3353)

But it can also be a problem for tourism if nature has not been civilized. While the climate at Nahuel Huapi was praised as pleasant – also in comparison to the sultry summer in Buenos Aires – the Iguazú National Park unleashed a debate about tropicality (Driver and Martins 2005). The tropical imaginaries were created with the European colonial penetration of the regions around the equator belt, and can be found in Argentina in the subtropical rainforest of the *Mata atlântica* in the border region with Brazil and Paraguay. In the various reports on the journey to Iguazú in the mid-1930s, the emphasis was not only on the difficult infrastructure but also on the unhealthy climate. In combination with blood-sucking insects, the humid heat creates an unbearable environment for civilized people. Malaria in particular was becoming a serious problem in the zone. In October 1935 Julio Amarante wrote to Bustillo:

> Life in this place is hardly bearable at this time. So far the heat has not been excessive, but when summer comes we will suffer in the small wooden hut where we live. Now the mosquitoes, the Mabrigüys, are the most intolerable of all plagues. (Letter Julio Amarante to Bustillo, 31.10.1935, AGN, Legajo Bustillo, Caja 3344)

The hot and humid climate and the plague of mosquitoes are a particular cause of concern for the parkers, as they might bring tourism to a standstill. For Amarante, the solution to the problem was to implement civilization measures, such as destroying weeds, clearing forests near settlements, using flamethrowers to combat mosquitoes, building sewers and draining swamps (Letter Julio Amarante to Bustillo, 31 Oct 1935, AGN, Legajo Bustillo, Caja 3344). However, despite subtropical conditions and temporary mosquito plagues – such as in the austral summer of 2016, when a dengue epidemic hit the Brazilian-Argentine border area – the Iguazú waterfalls remain an international tourist attraction.

Despite the general economic and geopolitical orientation of the national parks, however, the nature conservation argumentation of Bustillo is not purely benefit-oriented. In a letter to the Minister of Agriculture, he highlights the need for major investments in the USA, which would also bring monetary returns. But the national parks also had advantages that go beyond the material as Bustillo writes, "it is not possible to estimate it in monetary terms: recreation, education, health, affirmation of patriotism" (Letter Bustillo to Minister of Agriculture Luis Duhau, February 1935, AGN, Legajo Bustillo, Caja 3344).

Exequiel Bustillo can be credited with transferring the efforts to institutionalize national parks that have existed since 1903 into national legislation as well as establishing the functioning national park authority, DPN, which was later renamed Administración de Parques Nacionales (APN, National Park Administration). He was able to fall back on a team that, like Emilio Frey, had been working on the national park issue for decades. Unlike academic and civil society institutions such as the Sociedad Forestal Argentina, the Amigos del Árbol, or the Sociedad Argentina de Ciencias Naturales, Bustillo was not driven by the idea of nature conservation. He pursued a geopolitical interest. He even explicitly dismisses the later efforts to protect nature that began in the 1940s and culminated in the 1960s as useless when he writes:

> It is about making a nationalist effort in the border regions (a fact which my successors in the office have thrown quite in the dust with their conservationist theories). (Bustillo 1997: 107)

Bustillo, on the other hand, was driven by the idea of colonization through tourism. Although Bustillo had profited fundamentally from the pseudo-democratic policy in the *década infame*, relations with the Argentine government deteriorated in the early 1940s. Under the short government of Ramón S. Castillo (1942-1943), which ended with the military coup of 1943, the conservative oligarch Bustillo submitted a letter of resignation. In this letter of 16 September 1942 to the Minister of Agriculture, Daniel Amadeo y Videla, Bustillo justified his move with the massive underfunding of the National Park Authority (Letter Bustillo to Amadeo y Videla, 16.9.1942, AGN, Legajo Bustillo, Caja 3353).

The financial debate was of such magnitude that even the complete cancellation of the budget for the National Park Authority was debated (Letter Bustillo to the Diputado Nacional Ismael Lopez Merino, 12.9. 1942, AGN, Legajo Bustillo, Caja 3353). While Bustillo emphasizes that the expenses of the authority have been very moderate and at the same time the financial requirements have increased due to the newly established parks, the newspaper *La Prensa* points in a May 1942 article to economic problems in the tourism business. *La Prensa* attributes the deficit of 30 percent of the budget mainly to the lack of income from travel taxes. This is explained by the significant decline in international tourists – probably due to the Second World War. While there were still 1,210,301 pesos of income in this category in 1938, in 1941 it fell to 413,673 pesos, almost a third of this figure. Furthermore, *La Prensa* also shares Bustillo's assessment and argues that in view of a massive expansion of the park area – by 2,000,000 hectares in recent years alone – a budget of 2,500,000 pesos "does not seem to be sufficient to cope with the execution of the necessary new works and to maintain the existing ones" (*La Prensa*, 26.5.1942).

In the end, Bustillo does not resign in this situation, but the end of the Bustillo economy in the national park movement was foreseeable. On June 4, 1943, a group of fascist-oriented officers, the Grupo de Oficiales Unidos, to which also the later President Juan Domingo Perón belonged, organized a coup against the conservative government. Unlike the military government under Justo, the new military dictatorship was no longer under the control of the oligarchic elite. The political influence that Bustillo had enjoyed under the strictly conservative governments was thus lost. Immediately after the coup, Bustillo resigned on June 8, 1943 (Bustillo 1997: 494). He remained in office for several months. But his days were numbered. On 19 May 1944 Exequiel Bustillo finally resigned from the management of the National Park Authority.

In view of the political crisis in Argentina, the end of the existing national park policy was palpable at the beginning of the 1940s. And so, it might almost sound like a premature eulogy on Bustillo's work when *La Prensa* takes stock of seven years of national park law. With an allusion to the efforts of Bailey Willis, the newspaper acknowledged – in the spirit of Bustillo – in particular the successful tourist development of northern Patagonia:

> Bariloche is no longer the remote place accessible only to a small core of wealthy, risky hikers or foreigners who love a landscape similar to that of their land of origin; today Bariloche is visited by thousands of people, and we are sure that the influx of tourists in the coming years will add up to an overwhelming magnitude, which is particularly pleasing. (*La Prensa*, 26.5.1942)

While in the press the national park policy under Bustillo was rated highly positive with regard to the development of the remote provinces and the promotion of tourism, even after his final resignation in 1944 there was a critical public debate about Bustillo. An unknown defender of Bustillo, who does not fully share Bustillo's approach, was compelled in the face of this situation to reject the less "civilized" attacks and to restore Bustillo's "personal prestige". This was probably a distant friend of Bustillo's, who worked as a businessman in Patagonia. One of the main points in the debate about Bustillo is his biography. "It has been said that Dr. Bustillo's views correspond to his oligarchic education aimed at preserving the dominance of the coastal provinces over the interior". Here we are dealing with a double critique, which on the one hand is based on the oligarchic position, and on the other hand expresses the concisely formulated contrast between Buenos Aires and the regions – already formulated by Sarmiento in the *Facundo*. In defense of Bustillo, the unknown author refers to Bustillo's selfless and self-sacrificing work for the fatherland. At the same time, he emphasizes Bustillo's unique combination of visionary creativity and pragmatic assertiveness with regard to the national parks when stating, "His extraordinary dynamism created these true fantasies that were valid for his epoch. He constantly travelled the Patagonian streets and lived with the settlers there, whose problems he recognized". In addition, the author praises the efficient administration of the national parks, which often established state authorities for the first time in remote areas. The restrictions of the national park law, in contrast, were not caused by Bustillo, but by the legislation (Unknown author, "El doctor Bustillo y la Patagonia", without year, probably around 1946, AGN, Legajo Bustillo, Caja 3348).

But criticism cannot be wiped away that easily. Because even in the direct vicinity of Bustillo there was resentment. A member of the National Park Commission lists Emilio Frey, a historical figure for the development of northern Patagonia and the establishment of the Nahuel Huapi National Park, in view of the crisis of the national park authority, as also listing some mistakes:

> Firstly, the Commission, which is present in Buenos Aires and of which I am a member together with Pinedo, Bustillo, Dodero, Larriviere, Ortiz Basualdo, etc., cannot carry out its tasks effectively because they are very busy people who do not attend meetings. (Letter of commissioner Parque Nacionales to Frey, 16. November, year unknown, AGN, Legajo Bustillo, Caja 3347)

In view of this problematic situation, the member of the National Park Commission advocated a fundamental renewal and rejuvenation of the Commission. At the same time, centralism was to be overcome to the extent that not all meetings would take place in Buenos Aires, rather in the provinces where the parks are located. This internal critical statement makes it clear that the oligarchic, charitable model of the National Park Commission came to its political end. The combination of an elitist class with exclusive access to the political field and a focus on the undisputed national center of Buenos Aires was no longer functional for national park management that intervened in regionally different provinces and also had to meet scientific demands.

In addition, a generational upheaval was inevitable. Under Bustillo there was great continuity within the Commission, but no significant renewal. With Bustillo's resignation not only did the director of the Parques Nacionales resign, but also the founding generation of the National Park Authority.

Bustillo himself was never able to make friends with this concept of a national park that was focused on nature conservation. The lack of environmental ethics, which is also affective, is perhaps better expressed in the following anecdote than in many a pamphlet. Bustillo obviously did not care much about the protection of the fauna. Instead, his relationship to the fauna fits into an aristocratic context for which big game or predator hunting contributed to status. However, while U.S. President Theodore Roosevelt himself indulged in big game hunting, Bustillo was solely concerned with trophies. In 1937 he commissions his director in the Iguazú National Park, Julio Amarante, to get him a leopard skin. This then caused Amarante trouble, as the trade with leopard skins was forbidden. He informs Bustillo that the leopards themselves have become rare in the national park when he writes:

> These little animals are no longer abundant in the area, although they still exist, but like the people from whom they must have learned something, they have an incredible attachment to their own skin and do not like to part from it. (Letter Julio Amarante to Bustillo, 13.8.1937, AGN, Legajo Bustillo, Caja 3345)

However, Amarante finally managed to get hold of a small but fine leopard skin, which he then intends to send together with a larger fur still to be obtained. He is well aware of the legal difficulties, especially with the shipment of the delicate goods as he notes:

> There is some inconvenience with sending the remains of these critters, which tend to defend the lives of those still in our area, but I believe they can be sent by postal order without danger. (Letter Julio Amarante to Bustillo, 13.8.1937, AGN, Legajo Bustillo, Caja 3345)

A month later, Julio Amarante announces the good news that he has found a large leopard skin in Brazil. The risk of the illegal shipment of the fur seems to have kept him busy. The solution he finds does not lack a certain irony. The safest way to avoid the increasing controls for the protection of species is to send the fur to the DPN, which thus becomes the transshipment point for the illegal fur trade:

> After I had thought about it again, I decided to send it in official function by order to the national park management. I think it is the only way that they are not endangered, because, as I told you, it is quite strict to control the possession of these remains. (Letter Julio Amarante to Bustillo, 16.9.1937, AGN, Legajo Bustillo, Caja 3345)

The fact that Bustillo did not remove these delicate letters from his private archives before handing them over to a public institution like the

General Archive of the Nation makes it clear that the question of offences against nature conservation did not cause him any loss of prestige even years later.

In his memoirs Bustillo makes an attempt at the theoretical classification of the Argentine national park model under his aegis. He makes an ideal distinction between three paradigms of the national park movement. The first he calls the "orthodox tendency". It consists of placing an area under state control for the protection of flora, fauna and landscape and restricting or prohibiting any form of economic use (Bustillo 1997: 363). This model failed for him, however, as he shows in the emblematic case of the Yellowstone National Park:

> It has been made artificial to such an extent that contact with nature has become a great dream. The camps of indigenous people and trained bears sound more like a theatre or circus than an authentic picture of the natural wilderness. (Bustillo 1997: 364)

Ultimately, an anti-imperialist undertone also resonates in the criticism of this tendency when he insinuates that the park manager Gustavo Eppens is infected was infected with "orthodox virus" because he had "his head full of Yanqui literature" (1997: 372). This so-called orthodox tendency is juxtaposed to an "eclectic tendency," which – following Bustillo – can be defined as follows:

> Those who do not abandon the cult of nature and the preservation of the authentic untouched landscape, while not neglecting other necessities of the nation or, let alone, sacrificing them to the fanatical cult of a simple dogma. (Bustillo 1997: 369)

Contrary to his own claim, this eclectic tendency is not really positively defined, but is based on the differentiation from the "orthodox tendency". Bustillo sees himself as the main representative of this eclectic direction and inscribes himself in a line of ancestors from Francisco P. Moreno to Carlos Thays to Bailey Willis. Thus, an undertone resonates that marks this tendency as a genuine national, Argentine national park policy.

The third tendency that Bustillo sees in the debate is the negative one. This stance is completely against the designation of nature reserves, as the nation must use natural resources and cannot afford the luxury of designating unproductive areas (1997: 373). The polemic with which

Bustillo wrote these rather non-analytical passages on national park policy was hardly necessary in the nine years in which he headed the national park authority. The supposedly "orthodox" voices, as they might have existed under the Sociedad Forestal, the Amigos del Árbol or the Sociedad Argentina de Ciencias Naturales, were clearly silenced and are not mentioned by Bustillo himself anywhere – neither in the memoirs nor in his letters. In this respect, the polemic is more likely directed against his successors, who from the 1950s onwards were increasingly oriented towards nature conservation.

Exequiel Bustillo with militaries in North Patagonia
(AGN, Legajo Bustillo, Caja 3354).

Argentine National Parks in a Transnational Perspective

> Argentina liked to compare its role in world history with that of the United States of the North, and in reality, despite the differences resulting from the disparity in the numbers of the inhabitants, there was a certain analogy between the development of the two nations. (Reclus 1894: 594)

What the anarchist geographer Élisée Reclus writes here in the 19th volume of his mammoth work *Geographie universelle* about the role of Argentina in world history also applies to its pioneering function in environmental history in the conception and establishment of national parks. Both societies were settler societies, which at the end of the 19th century were faced with the Herculean task of integrating areas recently conquered by the indigenous peoples living there into the national territory and of materially and imaginarily civilizing nature there. Against the background of this homologous initial situation, the advocates of the national park idea repeatedly resorted to the example of the USA in various situations. Nonetheless, it could also be shown that this is not a simple transfer model. With the exception of the U.S. geologist Bailey Willis, no national park activist from North America intervened directly in the Argentinean debates. And every conception of the U.S.-American idea was not copied, but adapted and modified. In this sense, the idea of wilderness, which is very influential in the USA and is also influenced by North American transcendentalism, could not prevail in Argentina. The idea of civilization, domestication and colonization of the peripheral regions was too strong here. It could also be clearly shown that the Argentine national park debate took place in a transnational interweaving space, so that here – mediated via pivots such as the French-Belgian landscape architect Carlos Thays, the German botanist Carl Curt Hosseus and the Belgian botanist Lucién Hauman, to name only the most important, – international ideas and concepts circulated, which also had an effect on the Argentinean context and in turn influenced other contexts – for example the reception of Thays' idea of Argentine national parks in France.

A remarkable richness and a high level of reflection can be observed in the debate about national parks. The concepts of national parks that have prevailed in Argentina, also because of the natural environment of the country, have special characteristics that stand out from the transnational debates.

In view of the outstanding importance of the landscapes of the Iguazú Falls and Lake Nahuel Huapi, which had been defined as natural wonders, the importance of fauna appears to be of secondary importance in Argentine nature conservation discourse. Thus, the Argentine national park debate diverges completely from the debate in sub-Saharan Africa, which is marked by the importance of big game (Ford 2012, Gissibl, Höhler and Kupper 2012). The debate about nature conservation in North America has also been influenced to some extent by big game – especially bison – in contrast to predators – such as the puma – some of which have been released for hunting, while at the same time the protection of birds has been of outstanding importance. Internationally, the importance of fauna was particularly evident at the International Conference on Nature Conservation held in 1913.

In the aftermath of the passing of the National Park Act, however, there was some debate in Argentina about the protection of fauna. In a letter to Exequiel Bustillo, the president of the Comisión Nacional Protectora de la Fauna Sudamericana in Argentina explicitly distances himself from critical statements that the well-known zoologist Angel Cabrera Latorre must have made about Argentine national park policy (Letter from the President of the Comisión Nacional Protectora de la Fauna Sudamericana to Bustillo, 1.11.1936, AGN, Legajo Bustillo, Caja 3345). A letter written by Hugo Salomón – President of the Federación del Fauna Sudamericana – in the following year from Dalat, Indochina, to Bustillo clarifies the debate. Salomón underlines:

> The division of the animals into useful and harmful and the resulting permission to hunt wild boars and pumas is incompatible with the idea of providing future generations with a protected area on native soil in its natural state. (Letter Hugo Salomón to DPN, 26.11.1937, AGN, Legajo Bustillo, Caja 3345)

Against this background, Salomón calls on DPN to refrain from advertising hunting in the National Park in its brochures. He also refers to the reserves that the British have introduced in Malaysia, where hunting elephants and tigers is completely forbidden. In addition, he refers to the international debate context on nature conservation:

> As early as 1913, the Swiss Professor Sarasin, one of the fathers of the idea of nature parks, attacked the North American parks in a memorandum to the International Commission for the Protection of the World Nature and opposed the fact that they had not understood well what nature reserves were about. (Letter Hugo Salomón to DPN, 26.11.1937, AGN, Legajo Bustillo, Caja 3345)

This exchange of letters now also indicates the high degree of transnational integration in the national park debate. Against the dominant U.S.-American national park paradigm, Salomón would like to see the Argentinean parks oriented towards the experience of protecting fauna in the Asian colonies of Western Europe. In a letter to Bustillo and Lynch, Salomón then formulates the hope that the international experiences mentioned will be beneficial to the Argentine example (Hugo Salomón to Bustillo and Lynch, 25.11.1937, AGN, Legajo Bustillo, Caja 3345).

However, Exequiel Bustillo makes no secret of what he thinks of the president of the Federación de la Fauna Sudamericana and his ideas of nature conservation. In his memoirs he identifies Salomón as the main representative of a national park policy that he calls "orthodox". He writes:

> In our time here in Argentina, Dr. Hugo Salomón was the paladin of this school, a typical theoretical naturalist. He was president of the South American Fauna Foundation, a title that was as pompous as it was indicative of the non-working world in which this good doctor lived. (Bustillo 1997: 364)

Salomón must also have protested against the road construction in the national park, as it disturbed the fauna there, especially the birdlife. As a result, Bustillo, with a certain arrogance of power, described the personal conversation as follows:

> The conversation took place in my office, and I confess that I was so amazed at such a request that I looked at it with greater attention to make sure I was not talking to a madman. Of course, his request came in through one ear and out through the other. (Bustillo 1997: 364)

Another transnational national park concept focused on the protection and use of natural resources, especially forests. These ideas are linked both to the approach of sustainable forest management as developed in Prussia and absolutist France, and to the subsequent ideas of the U.S.

forestry expert Gifford Pinchot. Especially in neighbouring Chile, under the leadership of Federico Albert, such a concept for the protection of the forest was established. In this respect, an ecological debate strand of the time in Argentina also links the concept of "conservacionismo" with that of "reforestación" and pleads for a more integrated forest policy, which also tackles the issues of uncontrolled deforestation by unrecognised settlers and the danger of forest fires. Even though a unanimous declaration "with a tendency towards nature conservation and reforestation of our forests" was adopted at the 2nd Congreso de Municipalidades de los Territorios Nacionales, these approaches did not go far enough for the major landowner and representative of the elite, Lorenzo Amaya. With comments in the major Argentinean daily newspapers, he advocated the further implementation of policies to protect the forests. This makes it clear how strongly the idea of conservation was anchored in the conservative elite, even beyond the tourist variant pursued by Bustillo and the DPN.

While parts of the elite, for their part, emphasized a landscape-aesthetic concept of natural heritage that was intended for the moral edification of the nation, the forest representatives relied on utilitarian justification. Forestry engineer Otto Neumeyer, who was the first director of the Lanín National Park from 1937 to 1939, describes the wealth of forests in his National Park, which adjoins Nahuel Huapi:

> As you know, there are vast areas covered with forests of high commercial value, which represent a genuine privilege and the most valuable natural resource of this reserve, exploited according to criteria and with appropriate cultural measures, which will not only help to conserve this national heritage in the long term, but will significantly increase its value over the years in order to obtain the maximum yield from the areas of this reserve. (Letter Otto Neumeyer to Bustillo, 26.9.1940, AGN, Legajo Bustillo, Caja 3346)

The forest is a resource that represents an economic heritage of the nation. In the spirit of forestry sustainability, this heritage can also be preserved for future generations through planned and controlled use.

"From yesterday to today". DPN 1938a. *Parque Nacional de Nahuel-Huapi. Historia, tradiciones y etnología*. Buenos Aires.

However, a national park model was adopted which, based on the preservation of iconic landscapes, aimed at the development of peripheral regions for economic – above all tourist – use and their integration into the national imagination. The dominant national park discourse follows the historical narratives of colonization and the development of the frontier. In this respect, Exequiel Bustillo aptly describes the national parks as a "true instrument of colonization". In keeping with the progressive thinking of the time, the aim was to civilize and modernize the region through tourism. This idea is expressed in the following sequence of images from a 1938 National Park brochure, which expresses the development in the areas of transport, accommodation, overcoming rivers along the historical sequences from the "traditional" Mapuche society to the beginning settlement by European pioneers to the works of the National Park authority.

In the second part of this book, which now follows, the aim is therefore to work out the various dimensions of colonization through national parks and to understand the integral connection between the national park concept and the comprehensive project of colonialization.

Central elements of colonization through national parks also go back to discourses and practices before Bustillo, while other central elements also unfold their effect through the national park management of Bustillo to the present day. However, after the departure of Bustillo in 1943, the National Park Direction underwent significant conceptual changes, the processing of which would go beyond the scope of this book. For orientation, however, here are some key details.

Under the dictatorship established after the fascist military coup in 1943, the National Park Direction was not fundamentally harassed, but neither was it further developed. It was only with the rise of Peronism in 1946 that the National Park movement was to be upgraded and given a new direction. With the aim of integrating the popular masses into the nation, mass tourism in the national parks was promoted in particular. Only at the beginning of the 1950s did a national park policy explicitly geared to nature conservation come into being (Scarzanella 2002: 16-17). The main reasons for this strong focus on nature conservation were the increasing international debates and conventions. A first step in this direction was the Convention for Nature Protection and Wildlife Preservation in the Western Hemisphere of the Pan-American Union of 1940, which was signed by Argentina in 1941 and officially ratified five years later (OAS 1940). This convention, based on the US model, defines a categorization of nature reserves that represents a first step towards standardization in the Americas. Remarkably, the Spanish name of this Pan-American Convention on Nature Conservation is not a literal translation of the English name. Rather, the English title concentrates on the protection of flora and fauna, while the Spanish name explicitly refers to the protection of "natural landscape attractions."

In the 1960s, this trend towards international standardization was to increase further and ultimately lead to internationally binding standards. Central actors in this development of a global environmental protection regime (Meyer et al. 1997), which establishes worldwide "standardized blueprints of institutions and practices" (Gissibl, Höhler and Kupper 2012a: 4), are international non-governmental organizations such as the International Union for Conservation of Nature (IUCN). The IUCN was founded under the name International Union for the Protection of Nature on 5 October 1948 in Fontainebleau, France. Since 1962, it has organised the World Park Congresses, at which strategies for the development of nature reserves are coordinated in a transnational exchange. The international Red List of Endangered Animal and Plant Species, introduced in 1963, is one of the broad-based mechanisms of monitoring and has also led to global standardization in the field of biodiversity. In 1978, the IUCN introduced a system to make protected areas globally comparable according to a defined catalogue of criteria. In interplay with these trends towards internationalization, national governments implemented globally established norms and categorizations. The openness of meaning and regional diversity, which the concept of the national park thus still had at the beginning of the 20th century, was replaced by an internationally standardized "global environmental regime", which emphasizes not only the national imagination but also the importance of nature conservation in global contexts.

The National Park as an Instrument of Colonialization

On 18 July 1938, the Argentinean daily *La Nación* published an article entitled "Patagonia is an unknown region according to explorer Herbert Wilkins". *La Nación* comments on a contribution by the Australian explorer Sir Hubert – mistakenly renamed Herbert in the article – Wilkins. Wilkins was one of the most famous explorers of his time, mainly because of his aerial explorations of the Arctic with breathtaking photographs, as well as the first exploration of the eternal ice with the submarine 'Nautilus' – a clear allusion to Jules Verne's novel *20,000 Miles Under the Sea*. Wilkins is certainly one of the last personalities of the extinct genre of explorers. For as he emphasizes in the article, the "white spots" of the earth, whose exploration had inspired the imagination and ambition of the explorers, are now almost completely to be assigned to the surveyed and geographically recorded areas. Wilkins himself considers only Amazonia, Tibet, Northern Siberia, Greenland and the North Pole as well as, at the other extreme of the western hemisphere, Patagonia and the South Pole as still uncharted territories.

But exactly this assessment of the wilderness of Patagonia and the spreading of the "legend of the 'wild' Argentina" in Wilkins' contribution causes great indignation with the unknown journalist of *La Nación*. Because for him Patagonia is developed and civilized, there is nothing more to explore there. The white spots have long been filled: "In these latitudes there is nothing more to discover, even if a famous explorer claims the opposite, there are still people living in Patagonia who are susceptible to the fear of the presence of a civilized person". The Argentine journalist thus makes it clear that the colonial development of Patagonia, which was still controlled by indigenous peoples of the Mapuche and Tehuelche at the end of the 19th century and was hardly developed in terms of infrastructure at the beginning of the 20th century, was largely 'civilized' and integrated into the nation state in the geopolitical imagination of the Argentine elites just 50 years later. But how can this integration – or rather internal colonization – be explained?

"Patagonia is an unknown region according to explorer Herbert Wilkins". *La Nación*, 18.7.1938.

The origin of the newspaper article gives information. I found it in the estate of the Argentine politician, lawyer and pioneer of the state administration of the national parks Exequiel Bustillo. Bustillo would have emphatically and enthusiastically voted for the objection to Wilkins made here by the unknown journalist. For his political project as head of the National Park Authority consisted – as he expressly emphasizes in his memoirs *El despertar de Bariloche* – in completing the colonization of Patagonia. He sees an organic link between the establishment of national parks and the project of colonization:

> It is therefore only right to recognize that in the times of our management the national parks have not only safely stepped into the most beautiful corners of the Argentine soil, but have actually developed into a true instrument of colonization. (Bustillo 1997: 15)

Before the National Park Direction (DPN), however, it was the researchers, natural scientists and explorers who encouraged the colonization of the region. With their "discoveries" and the resulting renaming of topographical features, flora and fauna, they carried out a true Semantic colonization of the landscape. They were well aware that they were an integral part of a colonial expansion. Thus the zoologist Adolfo Doering gives his chapter on fauna in the Official Report on Patagonia of 1881 the title "General Observations of the Fauna of the Conquered Territory" (1881: 7). One of the last representatives of this social type is – similar to the discoverer Wilkins – the U.S.-American geologist Bailey Willis, who travelled North Patagonia on behalf of the Argentine government from 1910 to 1914. Willis sees his geological survey and classification of the lands according to their potential for agricultural and forestry use as the basis for a large-scale colonization project. Here he clearly sees the parallel to the land surveys in the west of the USA:

> This Argentine border had only recently been conquered and has not yet been fully brought under control. The sparse population consisted mainly of squatters who lived on the rich lands from their herds and gardens. Almost the entire area still belonged to the nation and was an asset that could be used to promote solid colonization, for which the uprising was to lay the foundation. (Willis 1947: 107)

With these dynamics, Argentina found itself in the midst of a tendency towards internal colonization and the closure of the frontier, which was to culminate and end at the end of the 19th century. For the end of the 19th century, the French geographer Élisée Reclus saw only two border regions in Argentina with worthwhile resources that still had to be fully developed:

> In the northeast the country of the missions, in the west the Andes valleys, where the rivers Colorado and Negro have their springs. In these regions of fertile soil, clean air and a wonderful climate, which alternate in the seasons that correspond to the temperament of the immigrants from Europe, there is room for millions of people. (Reclus 1894: 595)

In the northeast (Chaco, Formosa and Misiones) these are the areas that were incorporated at the end of the Paraguay War (1865-1870), while in the south they are the areas that were incorporated by General Julio

Argentino Roca after the conquest campaign against the Mapuche and Tehuelche in 1879 (Ruffini 2007). The so-called *Conquista del Desierto* was a bloody military campaign with genocidal tendencies, which resulted in the defeat of the Mapuche indigenous people in alliance with the Chilean army. While Patagonia was thus considered a dangerous territory due to the military conflict with the Mapuche and Tehuelche, which was now to be cleared of all indigenous traces, the situation in Misiones was different. Here, at the end of the 19th century, the idea of an "intercultural agreement, a pact of the jungle," according to which the indigenous population, which was presented as passive, tolerated the new settlers or even helped them, had increasingly prevailed (Wilde 2007: 92-93). This geopolitical imagination of the region replaced an idea of mission determined by images of an impenetrable, hostile jungle dominated by wild indigenous people. This change in the geopolitical imagination now indicates that the European settlers had lost their fear of the unknown region and had increasingly begun to open it up – even imaginarily.

A peculiarity of the state-building process in these peripheral regions, which were regarded as strongholds of chaos and unrest, is that between 1880 and 1950 they were incorporated into the nation-state as 'national territories,' but not as provinces, according to the constitution of 1862. On this basis, the nine government districts of Misiones, Chaco, Formosa, La Pampa, Chubut, Neuquén, Rio Negro, Santa Cruz and Tierra del Fuego were created. With a total area of 1,131,923 square kilometers they accounted for about 40% of the total area of Argentina (Ferrero and Pyke 2016, 2017). Despite this administrative regulation the question of the everyday enforcement of statehood, which was hardly given in these areas, arose. In this respect, the National Park Authority can be seen as one of the first state institutions to penetrate these remote areas.

Geopolitically, these inner border regions are also of particular relevance because at the same time they are interstate border regions. In the case of Misiones there are territorial conflicts with Brazil and Paraguay, while along the Andes chain the Chilean neighbour is identified as a rival – especially in Patagonia. It should be noted here that Lake Nahuel Huapi had been claimed by Chile even before the 1881 Boundary Commission ruling. Access to this region was also mainly via the Chilean side. In terms of population policy, it was particularly controversial that

Chilean Mapuche settled in this border region. After the military defeat of the Chilean Mapuche in 1881, the Chilean state introduced a reserve system between 1884 and 1924 that concentrated the Mapuche on 6.3% of their original territory. However, more than 40,000 Mapuche did not receive any land in the reserves (Kaltmeier 2004: 92-95). A part of this Mapuche then settled in the eastern Andes slopes of the later national parks Nahuel Huapi and Lanín. Their situation became increasingly precarious with the increasing state penetration of northern Patagonia. Here the state was committed to colonization and Argentinization. In order to deny these settlers access to land, their ethnic affiliation as Mapuche was made invisible. Before the conquest, the Mapuche had lived in Wallmapu, which included areas of the later nation states Chile and Argentina. Instead, they were referred to as Chileans, which meant that they were not actually on the Argentinean side. Social anthropologists Juan Carlos Radovich and Alejandro Balazote conclude, "[i]n this way the Mapuches have been doubly discriminated against, specifically as indigenous peoples on the one hand and generically as alleged Chileans on the other" (2009: 34).

Exequiel Bustillo was obsessed with the geopolitical rivalry with his Chilean neighbour. He saw the need to be in a "permanent state of alarm in the face of Chilean demands" to ensure the "defense of geographic integration threatened by a neighbor who, for geopolitical reasons, became more and more troubling day by day" (1997: 10). The Chilean Mapuche are a particular thorn in his side as he notes, "Every year hundreds of Chileans are seen crossing the border and settling down at will, without anyone stopping them and without our public forces being concerned" (1997: 13). As Bailey Willis had done before, they are held responsible for the forest fires in northern Patagonia. Bustillo, on the other hand, had no reservations about the German-Chilean settlers who had settled in Bariloche and who were engaged in intensive trade with their eastern neighbours. This mistrust of Chile is not Bustillo's personal fad, although it is particularly pronounced here; it is in keeping with the spirit of the times. Already in 1917, the experts of Patagonia and Argentine of the border commission, Francisco P. Moreno, warned against the Chileans:

> We know that Chile from the beginning of the border dispute strived for the line to run through the eastern foot of the Cordillera to Lake Nahuel Huapí. If it had succeeded, it would be a powerful, complete nation today, because it would have added to its strong brain and solid armor the indispensable: a well-fed stomach. (Moreno 1917)

And three years later, in a newspaper article, Chile's smallholder livestock industry is accused and the Argentine state is called upon to take control of forest resources:

> Chilean cattle breeding has invaded much of this area. From October 16th to Nahuel Huapí there are 1,500 miles of rich forests, such as Cypresses, araucaria, pine trees, beech trees, etc. These forests have been cleared and burned by foreigners in order to transform them into pastures where the inhabitants fatten their cattle [...]. These forests have trees up to 90 feet high, from which boards and wood can be extracted for all kinds of constructions. All this wealth, which has been mercilessly abandoned and destroyed, could return to the heritage of the state, and all the sovereignty rights, which we have somewhat neglected, would have to be exercised again in this region [...]. It will be a work of common sense and practical government to regain Argentine supremacy in these desert areas. ("El lago Nahuel Huapi debe ser un punto de turismo", newspaper clipping, MP, Colección Frey, Bibliorato 1)

The closing of the frontier in the high imperialism of the late 19th century had massive effects on the ecosystems. The importance of the turn of the century can also be clearly seen in the context of the debate about "great acceleration" in the Anthropocene (Steffen et al. 2015). While the geologists in the debates on the Anthropocene had agreed on the year 1950 as the global tipping point, known as the "golden spike," from which the man-made changes of the earth's surface outweigh the natural deformation processes, the century before is also highly relevant. In the 19th century the pace of deforestation accelerated worldwide. Between 1850 and 1920 as much virgin forest was lost worldwide as in the twice as long period from 1700 to 1850 (Williams 2006). From 1850 onwards a rapidly increasing loss of tropical virgin forests can be observed, while – almost mirror-inverted – the size of the cultivated area was increasing and almost reached its plateau value in 1950 (Steffen et al. 2015: 87).

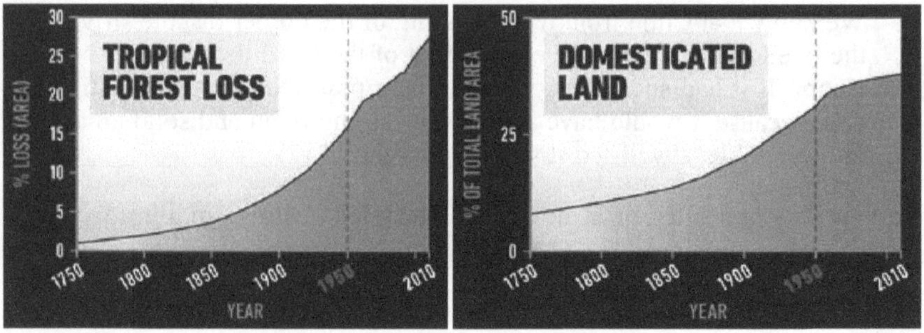

"Loss of tropical forests and increase in cultivated land" (anthropocene.info/great-acceleration.php).

The same applies to Argentina. In an environmental historical study on forest loss in Argentina, Adrián Zarrilli states that half of the forest has been destroyed since the beginning of the 20th century. This means in concrete terms:

> At the beginning of the 20th century, the forest area is estimated to be about one third of the land area or nearly 100 million hectares, while the latest information shows that there is currently only about 36 million hectares of forest area or about 13% of land, less than half of what existed at the beginning of the century, of which only about 15 million hectares of virgin forest and 800,000 hectares of planted forests would be economically viable. (Zarrilli 2008: 98)

In Misiones, the overexploitation is particularly massive, since the wood is of high quality and transport via the Paraná River to the capital Buenos Aires, where there was the greatest demand for wood, was relatively easy. For the region around Nahuel Huapi, Bailey Willis also states as early as the 1910s that primeval forests exist only in the areas of the Andean valleys on the Chilean border that are difficult to access:

> Only inaccessibility has protected this jungle from exploitation and fire. Now that they are made accessible, only the most constant vigilance of an energetic and efficient organization will be able to preserve them in the future. (1988: 384)

However, the establishment of such an organization, which Willis had in mind, has – apart from the national parks – failed everywhere. According to Zarrilli (2008: 98-99), unlike in many Latin American countries, the loss of forests is not due to direct land pressure, but to state failure in forest policy.

In addition to the loss of forests, there is an enormous loss of species worldwide and the associated dramatic decline in biodiversity, while at the same time the worldwide circulation of alien animal and plant species is increasing. The historian Charles Mann speaks in this context of the Homogenocene, i.e. the worldwide adaptation – or better homogenization – of the animal and plant kingdoms through a forced “columbian exchange”. Argentine national parks are not exempt from these processes either. In this context, they play the paradoxical role of on the one hand counteracting the effects of the anthropocene on deforestation and loss of biodiversity already known at the time with forest protection mechanisms, and on the other hand being one of the essential actors of biologic colonization with fish farming, tree nurseries and zoological facilities. In this sense, the concept of colonization used in this project not only refers to the confrontation with indigenous peoples and land conflicts, but also to the imaginary dimension and material transformation of the environment.

Agricultural Colonization and Land Conflicts

As in other regions of the world, Argentina at the end of the 19th century was concerned with eliminating the internal frontier. The areas in the north-east and south of the country that had been annexed in the 1880s had to be developed, penetrated by the state administration and modernized. The instrument initially preferred for this purpose was agricultural colonization. Ezequiel Ramos Mexía, Minister of Public Works from 1907 to 1913, was the mastermind behind the new agricultural colonization. In 1908 he designed the Ley de Fomento de los Territorios Nacionales to promote the integration and economic development of the peripheral national territories of Chaco, Formosa, Misiones and Patagonia. However, he turned away from the traditional etatistic-dirigist model of colonization, which he considered to be a complete failure (Ramos Mexía 1921: 12-14), and considered the task of the state to be, first and foremost, to the creation of the necessary infrastructure for colonization, particularly through the expansion of the railway (Bandieri 2009). Railway construction in northern Patagonia was intended to complete the desert conquest project begun by Roca, where he explains, "the conquest of the desert with weapons that you have carried out is not complete; we must add the conquest by rail, 'Argentineanize' Patagonia and use the Chaco River" (Ramos Mexía 1936: 205). Ramos Mexía was certainly also inspired by the rapid expansion of the railway in the western USA (Willis 1943: 1-2).

Ramos Mejia was also responsible for bringing the Hydrological Mission into the country under the leadership of Bailey Willis and supporting his projects for agricultural colonization and the establishment of an industrial city on Lake Nahuel Huapi. In a letter to Ezequiel Ramos Mexía, then Minister of Public Works, in 1913 Bailey Willis sketched out an extensive and detailed project of agricultural colonization for northern Patagonia, which also provided for the self-organization of the settlers in civil society in the form of trade unions. For its part, the state was to commit itself to expanding the transport infrastructure (Letter from Willis to Ramos Mexía, 6.2.1913, MP, Colección Frey, Bibliorato 2). But agricultural colonization was only one element of Willis' development plan for northern Patagonia. An industrial city as well as a national park were to be founded to accompany it.

What these geopolitical imaginaries of the agricultural colonies have in common is that they conceive of the corresponding areas as deserted. The indigenous population there is regarded as a remnant of times long past, which should be eliminated under the banner of universal progress and civilization. Accordingly, a newspaper article calls for the establishment of an Indian reserve at Lake Nahuel Huapi – similar to the situation in the USA or neighboring Chile. In this "colonia indígena," the indigenous people remaining in the region would be concentrated and placed under military supervision:

> The idea of concentrating the vagabond indigenous element in a colony with military garrison leads to the solution of the problem of the security of this region, to the civilization of the Indians and to the resistance against the degradations and irreparable damages caused by those who come from Chile, such as the forest fires and the looting of the areas so abandoned by the authorities, which are beautiful and fertile in wealth. ("Territorios del Sur", MP, Colección Frey, Bibliorato 7)

The author attributes a degree of intelligence to the Mapuche, but then argues with a degeneration, "Today, without leaders and government, they pursue raids and looting." And he continues, "They steal, kill and even ignite the magnificent state forests, which are the wealth and proud jewel of these regions." Faced with this scenario of threat and in the absence of state institutions, the unknown journalist welcomes the fact that landowner Jorge Newbery has been commissioned by the state to protect the forests ("Territorios del Sur", MP, Colección Frey, Bibliorato 7).

But the family of the big landowner Newbery pursues its own territorial interests in the region. The danger of the formation of large estates, which Ramos Mexía did not foresee in his settlement program, is obvious. Thus, in April 1919, Jorge Newbery's son Diego applied for grazing rights on 2,500 hectares of state-owned land at Traful Lake, which were connected to the Newbery's estate (Letter Diego Newbery to the Ministry of Agriculture, 29.4.1919, APN, Caja Permisos de Pastaje 1936, 1). But obviously this land was already occupied by Mapuche, so that land conflicts occurred. In 1923 a commission was sent to Lake Traful to investigate Newbery's search for land, which – as the commission report notes – had the peculiarity that the land applied for was located in the reserve established in 1908 and the National Park of the South finally established in 1922 (Informe Comisión Inspector del Neuquen, 2.5.1923, APN, Caja Permisos de Pastaje 1936, 1). The commission's work itself must have been seen positively by the Mapuche of the region, but after the commission left the region, fear returned.

And so, Elisa de Pimán, an elderly Mapuche widow who lives at Lake Traful, addresses a letter of complaint to the Minister of Agriculture. In the letter, she accuses the head of the family, Jorge Newbery, of making false accusations to the Ministry of Agriculture. Furthermore, in January 1919, his son Diego, together with local police forces, had forcibly expelled them:

> They set fire to our huts and we had to flee to the snow-covered hills where almost all our cattle died. (Letter Elisa C. De Pimán to Le Breton, 14.1.1924, APN, Caja Permisos de Pastaje 1936, 1)

In August 1920 Diego and Guillermo Newbery – also with the help of the local police – are said to have set the Mapuche family's huts on fire again. These acts remained unpunished and no compensation was paid. The widow also makes clear another strategy of expulsion. Newbery – so the accusation – would have achieved that her payments for the pasture rights were not collected by the state authorities, in order to thus have an excuse for the expulsion. For this reason de Pimán stresses "we want to pay, just like all settlers" (Letter Elisa C. De Pimán to Le Breton, 14.1.1924, APN, Caja Permisos de Pastaje 1936, 1). With this letter to the Minister of Agriculture, Tomás Le Breton, the widow tries to lead the central state power against the quasi sovereign power of the local landowner and asks the state for "his protection against Mr. Newbery and against all rich gentlemen who want to take away the state lands from us". She proposes the drawing of a border fence, and then speaks of their fears in the letter:

> We are afraid, Minister, that the Newberys or the police will make us victims again and that our years of work will be destroyed. That is terrible in our poverty and with so many small children, Lord. If they throw us out, where will we go? (Letter Elisa C. De Pimán to Le Breton, 14.1.1924, APN, Caja Permisos de Pastaje 1936, 1)

Unfortunately, the exact outcome of the conflict cannot be reconstructed from the available sources. But obviously Elisa de Pimán was able to defend the pasture rights successfully. In 1947 she became the target of renewed efforts by the National Park Direction, which had been established for more than 10 years, to expel her from their land within the limits of the park.

The fear of the Mapuche settlers at Traful Lake was by no means unfounded in the 1920s and 1930s. In November 1934, the shores of the lake, which attracts anglers worldwide, again became the scene of a violent conflict. The brothers Diego and Herodes Vazquez, both of Chilean descent, were accused of illegally occupying a state piece of land and were therefore ordered to evacuate. They resisted eviction and were shot dead by police. The "bad background" of the two Chileans is emphasized in the report about this fatal incident and the suspicion is expressed that both were smugglers (Letter Villarroel to Rodolfo del Castillo, 12.11.1934, AGN, Legajo Bustillo, Caja 3344).

While the land conflicts in the early phase of the National Park of the South, which was founded in 1922, focused on private actors, later the importance of the DPN in the expulsion of Mapuche and other settlers from the National Park increased from 1934 onwards. With increasing institutionalization, the National Park Direction also assumed an increasingly important role in resolving land conflicts in the National Park that were carried out between other actors in the park. In land conflicts between European settlers and Chilean Mapuche, the national park authority always seems to side with the settlers. In 1934, the year the authority was founded, Florinda Hinai, Millaqueo's widow, accused the Belgian settler Benito Vereertbrugghen of fencing in her property on the edge of Lake Mascardi, which she had been managing since 1920. This makes a pasture she owns in this area an enclave and denies her the right of way, which is also needed to drive the cattle to this very pasture. Due to the border situation, the Chilean consul also intervened in this case, which finally ended with compensation payments from Vereertbrugghen to de Millaqueo – probably through the mediation of park director Emilio Frey. The controversial pasture of Vereertbrugghen was thus closed (Report Francisco Sanjuan, 7.4.1934; Letter to Bustillo, 16.10.1931, APN, Caja 1934).

Founded in 1934, the National Park Direction quickly established itself as the most relevant and present state institution in the region. The National Park is thus the starting point of a veritable process of state-building in the peripheral areas of the nation. To this end, DPN systematically began to collect data on the settlers living in the Nahuel Huapi National Park (Staropoli 2010: 12-13). For this purpose, personal files were created that included the full name, the family situation, police entries as well as data on the cultivated parcel, such as the start of use, size and location, livestock and agricultural use. On this basis, the precarious rights of use, the *permisos precarios de pastaje* (PPOP, Precarious Grazing Permissions), were also issued and it was determined whether the fees fixed for use had been paid on time. The PPOPs also served as a standard that determined permitted and prohibited behavior. This data was used to standardize the population and led to disciplinary measures ranging from the imposition of fines to eviction. A report by the director of the Los Alerces National Park, which was in the process of being established, identifies "bad social elements" in the National Park that are already recorded in the documents of the authority:

> We have already recorded their background in the census. Moreover, and as soon as you consider it opportune and appropriate, shall I ask in writing for the removal of the bad element? (Letter to Bustillo, 26.9.1937, AGN, Legajo Bustillo, Caja 3345)

The visibility and authority of the National Park Direction was also increased by the introduction of uniforms for the park rangers. The director of the Nahuel Huapi National Park, Emilio Frey, describes the function of the uniforms as follows:

> It is almost unnecessary to note the advantages that would be made with such a point: more authority, more hierarchy, etc., as well as the flawless impression the tourist would have in view of a well uniformed employee. (Letter from Frey to Bustillo, 15.9.1936, AGN, Legajo Bustillo, Caja 3344)

"Parkranger" (Kirchhoff 1945).

Obviously, the establishment of the National Park Direction also triggered new activities in other state units. In 1935, for example, the Dirección de Tierras of the Ministry of Agriculture enacted a law regulating the treatment of settlers in state countries with forest stands. For example, slash-and-burn was banned. In addition, the rights of use were reviewed, thus laying the foundation for the evacuation of illegal settlers. In addition, a forest police force was established to implement the decisions on forest protection (*La Prensa*, 21.11.1935).

The population policy interventions of the National Park Direction were initially directed particularly against the Chilean Mapuche group, who had a vulnerable status and were also exposed to racist discrimination by society as a whole. This group was attributed a new identity as the "Chilean Mapuche" in the border region, which was marked by structural violence. In the reports of the Dirección de Tierras y Colonias – so argues the social anthropologist Florencia Trentini – the categories "indio" or "chileno" or their combination were already used "to reject any demand for land use, since they were regarded as 'lazy,' 'harmful,' 'drunk' and 'delinquent,' This legitimized the notion that these groups could not have access to public land as owners but could only be concessionaires or occupiers" (2015: 93). The Mapuche, however, contrast this external attribution with another picture:

> We are of indigenous origin, but we like to work and want to move forward. We want to make the fields work for our children by teaching them to love their homeland as good Argentinians. (Letter Elisa C. de Pimán to Le Breton, 14.1.1924, APN, Caja Permisos de Pastaje 1936, 1)

In 1937, the DNP director, Exequiel Bustillo, and the successor of Emilio Frey in the management of the Nahuel Huapi National Park, Alexis Christensen, had a serious conflict over the evictions. As already described, Bustillo was obsessed with the guiding biopolitical idea of pushing back the Chilean settlers and argentinizing northern Patagonia through population policy measures. In this sense, he authorized the evacuation of a Chilean settler without further consultation. For Christensen, on the other hand, the eviction was inadmissible and – with unusual clarity – he called on Bustillo:

> I would ask you never again to sign an eviction order without first hearing the person concerned in order to avoid these cases, which stir up every conscious being. (Letter Christensen to Bustillo, 5.4.1937, AGN, Legajo Bustillo, 3345)

Then Christensen continues and dismantles the nationalist-racist geopolitical imaginary of the "Chilean danger:"

> That you do not sell land to foreigners from the neighbouring country is fine when it comes to LARGE EXCESS, but this cannot be applied to small plots of land in the village. (Letter Christensen to Bustillo, 5.4. 1937, AGN, Legajo Bustillo, 3345)

And he continues to argue that there can be no talk of a Chilean strategy of expansion here:

> I do not see the slightest indication that there is any intention of penetrating these places economically and sending the Chileans here. In fact, the opposite is happening, because the government is trying to prevent people with high consular fees from coming for passports. (Letter Christensen to Bustillo, 5.4.1937, AGN, Legajo Bustillo, Caja 3345)

In addition, the descendants of the settlers already clearly identify with Argentina in the first generation. In conclusion Christensen then brings even more sharpness into the letter and warns against the use of Nazi methods in the national parks, a false nationalism and an unfounded fear of the Chilean neighbour should be the reason:

> that we commit injustices or systematic persecution of poor people and implement methods of aggressive nationalism with the devastating result as we have seen put into practice in Europe. (Letter Christensen to Bustillo, 5.4.1937, AGN, Legajo Bustillo, 3345)

In this sense, he calls on DPN in general – and Bustillo in particular – not to lose "nobility". A few months later, this dispute about the treatment of the Chilean settlers continues. In another letter to Bustillo, Christensen demands neutral and equal treatment for all settlers who are potentially evicted. Specifically, he refers to the unequal treatment of two settlers, which is obviously based on personal relationships and ethnicity. And he demands:

> They are not supposed to tell me that they are different people [...] I have noticed little difference between the spirit of the Swiss-Germans we have here and that of the other poor Chilotes who are to be expelled. (Letter Christensen to Bustillo, 26.10.1937, AGN, Legajo Bustillo, 3344)

Thus, the Danish-born Christian refers above all to the strong power of the German-Swiss group within the region, which is tolerated – or even promoted – by the park administration, while the Chilean Mapuche are regarded as undesirable segments of the population and invaders who should represent a geopolitical threat to Argentina. Christensen gives consideration to this position, which Bustillo has decisively advanced:

> As far as Argentinism is concerned, I think it is enough to mention that efforts are being made to establish a German school in Bariloche! The others, on the other hand, do not despise the Argentine school. (Letter from Christensen to Bustillo, 26.10.1937, AGN, Legajo Bustillo, 3344)

We do not know how Bustillo reacted to this criticism. But it is to be suspected that it did not completely fade away without consequences, since Bustillo appreciated Christensen very much. In his memoirs he describes him as "an efficient and faithful collaborator" and as "an exemplary civil servant who had what it takes to be a minister because of his correctness and diligence in office" (1997: 241). Although Bustillo did not change his basic attitude, it seems that in the national parks of Patagonia from 1938 to 1943 less land conflicts can be observed. Christensen's warning against an increasingly unwilling German congregation in Bariloche, however, should prove to be justified. A few years later, uniformed German Nazis marched through Bariloche as soldiers in large marches, and the school in question proved to be a refuge for refugee war criminals, such as SS officer Erich Priebke, who until the 1990s was still the chairman of the supporting association of the German school.

Around the years 1936 and 1937, however, criticism of the expulsions of Chilean settlers not only aroused within the National Park Authority, but also diplomatic disgruntlement ensued when dealing with the Chileans. During his visit to Bariloche in 1936, the Chilean consul expressed his concern about the evacuation of Chilean settlers and their expulsion to Chile. And in a handwritten side note to this report it was also noted that the complaint of the Chilean diplomat was well-founded, since there were two to three cases of unjustified evictions (Letter from Frey to Bustillo, 12.11.1936, AGN, Legajo Bustillo, 3345). However, the mood in Chile towards the Mapuche was no less racist than in Argentina. Accordingly, the Chilean authorities only half-heartedly supported the Chilean Mapuche in the national parks, or they even showed open rejection. The director of the Los Alerces National Park reports that the Chilean consul to him declared "that these people are bad here and in their country". The director of Los Alerces – in opposition to Christensen – takes this as an opportunity to point out to Bustillo:

> I believe, Doctor, that we must remain strong in order to eliminate all negative elements [...]. These people are not only sluggish, but also irre-

> sponsible and even bad. [...] Their incurable statism, their primitivism and this reluctance to do anything that belongs to us, which they keep for themselves, identifies them as unrecommendable social products. (Letter Intendant PN Los Alerces to Bustillo, 17.8.1938, AGN, Legajo Bustillo, 3345)

And he concludes with the economic argument of resource control:

> If we have completely separated ourselves from this type of settlers, our forest wealth will be better secured. (Letter Intendant PN Los Alerces to Bustillo, 17.8.1938, AGN, Legajo Bustillo, 3345)

Around the year 1947 a new wave of the expulsion of Chilean Mapuche from the national parks of northern Patagonia can be noticed. For this purpose, a list of the affected persons, most of whom are very ancient, is compiled. These include the widow Elisa C. de Pimán at Lake Traful and the Chilean Mapuche Juan Mora. Mora is married to the Mapuche Maria Dolores Inalef and settled on Lake Llum near Lake Mascardi in the 1920s. This renewed wave of expulsion also provoked criticism within the authority. The then director of the Nahuel Huapi National Park, Mignacco, comes to the following conclusion:

> In general, and according to what has been said, it is the opinion of the undersigned that, for elementary reasons of humanity, it would be appropriate to reconsider the measures to combat the evictions in question, since the majority of these settlers are elderly people who are absolutely lacking in resources and have numerous descendants. (Letter Mignacco to Oscar Arce, 13.5.1947, APN, Caja 1937)

The director of the Nahuel Huapi National Park also argues that these people have only non-transferable, precarious rights of use, which will therefore expire in the future. But for the descendants, whom he expressly calls Argentineans, the state must then provide sufficiently large replacement areas.

We do not know the answer from the headquarters in Buenos Aires, but a month later Mignacco is forced to clarify the situation in a letter to the Administrador General de Parques Nacionales. Long and wide he then explains that the Chilean Mapuche settling in the park are unwanted persons who are completely uneducated and uncultivated and can only lead a reasonably orderly life due to the supervision exercised by the park administration. And yet they indulge in alcohol, are responsible for forest fires and cause trouble. They lack any moral education, which also leads to incest and promiscuity (Mignacco to Irusta, June 7, 1947, APN, Caja 1937). In this letter, the Park Director thus cites almost every common racist prejudice, but without explicitly mentioning ethnicity. The humanistic argument based on compassion cited in the first letter is no longer to be found in this letter. Instead, it is now strategically weighed so that the effort that an expulsion would entail is not justified in view of the marginal, small parcels of land affected.

But even these arguments obviously do not reach the central park administration in Buenos Aires, and so on 6 September 1947 a letter is issued to Juan Mora in which the precarious pasture rights are withdrawn from him. In 1952 a new eviction notice follows. And in 1956 the eviction was probably carried out with the help of police forces, who had requested the park rangers for support. Despite this eviction, Mora continued to use the Lago Llum region as pasture for his 30 cattle in 1960. He himself had settled on the adjacent plot of land of his deceased wife. Thereupon, the park rangers intervene again, catching the animals and confiscating them. The costs of this action, amounting to 3,652 pesos, were charged to Mora, who had occupied "clandestine" land as an "intruder". Juan Mora was 70 years old at that time (Various sources, APN, 1937).

Nevertheless, not all Mapuche living in the Nahuel Huapi National Park could be displaced. Particularly in the 2000s, new organizational processes began in Bariloche, and to a lesser extent in Villa La Angustura and El Bolsón as well, in a growing indigenous Mapuche movement (Valverde 2011: 77; Policastro and Marks 2008). To this day, the land conflicts associated with colonization and the founding of the park are still violent. On 27 November 2017, during the evacuation of a land occupation on Lake Mascardi, the Special Command "Albatros" of the Argentine Navy shot 22-year-old Mapuche Rafael Nahuel. Beyond this sad, incisive event, which triggered national protests, the organization process has also been successful for the Mapuche who live in the park. The Comunidad Wiritray at Lake Mascardi was the first indigenous community to be officially recognised by the Instituto Nacional de Asuntos Indígenas inside the Nahuel Huapi Park in 2000. The comunidad mapuche Wiritray was founded in 1998 by the families Treuque,

Rauque and Inalef – the latter was involved in the land conflict depicted around Juan Mora – who all owned PPOP. For a long time, Wiritray was the only indigenous community recognized by the National Park Administration inside Nahuel Huapi. But over the next 10 years, more indigenous communities were founded, four of which are now recognized. There were also other occupations for land reclamation, while the park administration tried to implement new concepts of comanejo with the indigenous communities in the park (Trentini 2011, 2015). The fact that these land conflicts are state land in the national park and that the state, as the main actor, has to react politically to the conflicts, certainly contributes to the fact that the question of recognizing the rights of the Mapuche is on the political agenda in this region.

The expulsions in the national parks of Patagonia in the first decades of the foundation were mainly directed against the impoverished Chilean Mapuche. Nevertheless, even at the beginning of the national park's establishment, the European and Argentinean settlers in the region were concerned that they might lose their land rights. The basic conflict is that the idea of a national park is based on state ownership of land, while the settlers have already acquired private land rights. The newspaper *La Nación* wrote in a long essay about the "Parque Nacional de Nahuel Huapi" in 1914:

> With the settlement of the same regions and the fact that these settlers acquired rights as owners under the Colonization Act or the titles granted by the Ministry of Agriculture, it became increasingly difficult to implement the Moreno Project. (*La Nación*, 17.7.1914)

On the occasion of the visit of the former President of the United States of America, Theodore Roosevelt, a confessing advocate of the national park idea, to Bariloche in 1913, the settlers addressed the founding father of the Nahuel Huapi National Park, Francisco Pascasio Moreno, and expressed their concern as follows:

> We address ourselves especially to you, Perito Moreno, who in noble generosity has offered to the Argentine people, the well-deserved fruit of many years of intensive work and hard need, the National Park. We deserve the general applause which today is so unselfish. But also on this occasion – we can see spirits in broad daylight – it appears to us, on the basis of reports from the press and other rumours that the sword of Dam-

> ocles is above our heads: the danger of expropriation. (Memorial to Dr. Francisco Moreno and Dr. Isidoro Ruiz Moreno, 26.11.1913, MP, Colección Frey, Bibliorato 1)

Furthermore, the settlers here also refer to the state strategy of agricultural colonization, whose reputation they themselves have followed, so that the state now has to provide legal protection.

> Is this not what a colonizing government wants? Men who take root in their country like a tree? Men who love the soil and the land, be it their homeland or the new homeland of their children, who they will defend in an emergency with their blood and their lives? (Memorial to Dr. Francisco Moreno and Dr. Isidoro Ruiz Moreno, 26.11.1913, MP, Colección Frey, Bibliorato 1)

And then they explicitly turn against the U.S. model of wilderness, with deserted parks, and postulate the fundamental compatibility of agricultural colonization and tourism in the national park regime:

> There is no park without residents, without settlers, who make this region habitable even for those who are only looking for a short rest. (Memorial to Dr. Francisco Moreno and Dr. Isidoro Ruiz Moreno, 26.11.1913, MP, Colección Frey, Bibliorato 1)

In this sense, the intervention culminates in the demand to take the Damocles sword from their heads and immediately issue final land titles.

The settlers remain stubborn. In view of the new political situation after the radical president Hipólito Yrigoyen took office in 1916, they sent a memorandum to the president in the same year (Memorial al presidente de la república, 14.12.1916; MP, Colección Frey, Bibliorato 1). According to Francisco Moreno's son, this memorandum was drafted by Emilio Frey and his father. Thus Moreno explicitly has dissociated himself from the U.S.-American wilderness model, and now supports the limited agrarian colonization in the national park as well. Thus the appeal of the settlers from the year 1913 has arrived at the now aged Moreno, who is now committed to taking the sword of Damocles from the heads of the settlers. In the memorandum they explicitly refer to Willis' plans for the colonization of northern Patagonia and demand the expansion of the railway for tourism as well as the promotion of agricultural colonization along the lines of California. After this presentation of the geopolitical imagination of Northern Patagonia under the sign of agrarian colonization and tourism, the concrete request to the president to regulate land use – also in the national park – follows:

> As far as the Nahuel Huapi colony is concerned, we ask that all reserved parcels be made available to the pastoral colonization and that they be suitable for this purpose; also within the national park, the use of the land for dairy products should be allowed on suitable areas, as long as the forests are not destroyed. We want parks with life, the more people can live from the fruit of the land, the more wealth there will be for the region and for the land. (Memorial al presidente de la república, 14.12. 1916, MP, Colección Frey, Bibliorato 1)

In the regularization of land titles, however, the settlers explicitly oppose the expansion of large land holdings, so that even expropriation or at least parcelling out is demanded here.

While the National Park Direction left the European settlers in the park unharmed until the 1940s, in the 1960s the pressure on the settlers increased for reasons of nature conservation. The biopolitical strategy of the National Park Direction, which sought a new regulation of the relationship between population and territory in the park, is expressed as follows in a long letter of complaint from the Unión de Pobladores del Parque Nacional Los Alerces in 1966, which was probably made up primarily of settlers of European descent and Argentinean origin:

> With the establishment of the park, the National Park Administration considered all these inhabitants almost as invaders, and granted their residence permit a precarious character. It has thus subjected the settlers to the most severe restrictions; it has not allowed them to make improvements, increase their capital or make progress. In addition, each year they were imposed payments for the right to graze livestock, with heavy fines for those who exceeded the limit. (Letter Unión de Pobladores del Parque Nacional Los Alerces to the Governor of Chubut, 16.10.1966, BM-APN, Caja PN Los Alerces)

This leads to the following conclusion:

> The policy of the administration, which was officially recorded in some files, was specific: the depopulation of the park. (Letter Unión de Pobladores del Parque Nacional Los Alerces to the Governor of Chubut, 16.10.1966, BM-APN, Caja PN Los Alerces)

This resulted in a massive and long-lasting land use conflict, which lasted over 30 years in the case of the settlers in the Lanín National Park and led to a hardening of the fronts between settlers and DPN:

> And this is the reason for the conflict between the park management, which regards the residents as enemies – and treats them like enemies – and the residents, who do not go because they cannot go anywhere. (Letter Unión de Pobladores del Parque Nacional Los Alerces to the governor of Chubut, 16.10.1966, BM-APN, Caja PN Los Alerces)

But the settlers seem to have asserted themselves partially. While there were still about 100 settlers when the region was designated as a reserve with the aim of establishing a national park in 1937, today 30 settlers, who mainly live from cattle breeding, are still able to live in the correspondingly designated use zones of the Los Alerces National Park.

While the settlers in the Los Alerces National Park resisted the concept of a deserted park, other settlers have come to terms with the new situation. Thus Benito Vereertbrugghen accepted his eviction from the Nahuel Huapi National Park and is convinced of the nature conservation argument of the DPN: "[I]f the care of the park and its flora requires my retreat as well as that of all present and future settlers who are in the same conditions" (Letter Vereertbrugghen to DPN, 1.4.1969, APN, Caja 1936), then he accepted his resettlement without reluctance. While the respected settler Benito Vereertbrugghen, owner of the Hotel Tronador on Lake Mascardi and the Hosteria Pampa Linda, among others, could still be sure of the support of the National Park Authority in land disputes with the Mapuche widow de Millaqueo in the 1930s, the paper turns 35 years later. In 1969, the now very old Vereertbrugghen, who had lived in the region since 1915, was ordered to evacuate. He does not oppose the eviction itself, but asks for more respect and a delay, as 30 days are hardly enough to relocate his 500 cattle. This case is almost exemplary for the definitive end of the agricultural colonization around Lake Nahuel Huapi. And Benito Vereertbrugghen uses the request for eviction to bear witness once again to his pioneering achievement, which was no longer in demand in the 1960s. "I am a Patagonian settler who has inhabited and colonized border areas since 1915". He presents himself nostalgically as an exemplary pioneer:

> I have always adhered to the rules of the National Park, I have served in national defence (as captain appointed by the nation's armed forces), I have populated a vast mountainous area and have led it to success with my economic activities. All this, since the time when life in Patagonia was an adventure, an invaluable sacrifice, and this has justified in definition this great national hope that now awakens the Argentine conscience. (Letter Vereertbrugghen to DPN, 1.4.1969, APN, Caja 1936)

Obviously Vereertbrugghen was granted more time because of this intervention. The final evacuation was carried out in December 1975.

The tragic figure of Benito Vereertbrugghen shows the failure of the model of agricultural colonization and the end of the first phase of closing the frontier. In this respect, Vereertbrugghen presents himself as the last representative of the almost extinct social type of the robust pioneer to appear in the DPN files. Like the West in the USA, where the historian Frederick Turner set an academic monument with his essay "The Frontier in American History" after the Frontier was closed, the nostalgic memory of the Frontier in Argentina is now also being created. While Wild West Romanticism is booming in the USA, Exequiel Bustillo describes just as nostalgically the attitude to life of his foreman, who was "quite a settler of the Wild West" (1997: 56), who led the "settler life of a true colonist or 'pioneer,' with all the risks and inconveniences this entailed" (1997: 61). But then to come to the realization for himself that he himself prefers urban amenities with telegraphs, streets and shops.

But Vereertbrugghen was not just a simple pioneer. He also embodies the successful businessman of the Frontier, who contributed to the region's tourism by boosting the local hotel industry. Because of his strong social position as a member of the local elite, he was long regarded as untouchable in the park. In this sense, he was also able to rely on DPN's support in the land conflicts with Chilean Mapuche, who were generally criminalized as intruders in the park. At the beginning of the peak phase of Nahuel Huapi's tourist activities under the motto "Argentine Switzerland" at the end of the 1920s, there were no fundamental conflicts of interest between local livestock farming and tourism, as Vereertbrugghen demonstrated with his tourist installations at the foot of the Tronador. It was not until the international nature conservation discourse that increased in the 1960s that such personalities as Vereertbrugghen came under pressure and had to leave the park. In the process, an increasing bureaucratization and local disembedding of the Park administration and the APN can also be observed, so that personal merits and historically grown privileges had given way to the cold mechanics of the central bureaucracy. And so the eviction was carried out.

In the Iguazú National Park – in the extreme north-east of the country – the conflicts between agricultural colonization and the national park regime took on a different form than in northern Patagonia. There are two main structural differences to be noted. Firstly, the conflict with the indigenous population is less prominent, since the indigenous peoples there were conquered earlier – partly already in the 17th century – and for the most part missionized, while the remaining indigenous communities withdrew to remote regions of the jungle (Wilde 2007). Secondly, with the military colony already installed in the region, another strong state actor was represented in the region who pursued his own interests.

Against this backdrop, the newly founded National Park Direction and the military had a dispute over 50,000 hectares of land. The military had taken control of the Ayarragaray Hacienda, which had been purchased by the state in 1928 with the aim of establishing the national park. With the founding of DPN in 1934, the military was forced to reach this area of the new authority. However, the military was reluctant to hand over such a geopolitically important area to a new, hardly institutionalized authority. In addition, the hotel, which was located in the area, generated income that the War Ministry did not want to lose (Bustillo 1997: 428-429).

In July 1935, Exequiel Bustillo sent General Alonso Baldrich, a member of the DPN, to Iguazú to arrange this transfer of land, including the installations like a hotel (Freitas 2018: 115). On September 4, 1935, Baldrich Bustillo sent a detailed report about his trip to Iguazú. After a brief introduction about the inconveniences of the trip, he promptly talks about land use. In August, the President of the Republic, Military General Agustín Pedro Justo, issued a decree after the military had kept 20,000 of the 75,000 hectares of the big estate, while 55,000 hectares were transferred to DPN. Baldrich now suggests in his report that of these 55,000 hectares, 5,000 hectares of the DPN are perfectly sufficient to build hotel infrastructure, park management and a network of tourist routes. According to Baldrich, the remaining 50,000 hectares "have no characteristics that would justify their inclusion in the national park system". Instead, he proposes parceling the land to use it for a large-scale agrarian colonization project, which will involve the settlement of 2,000 families in the region (Report Baldrich to Bustillo 4.9.1935, APN, Caja 1935; Letter Baldrich to Bustillo, 3.10.1935, APN, Caja 1935).

At the beginning of October 1935 Baldrich then followed up in a further letter and takes – against parts of the DPN – a clear position against the incorporation of the disputed area into the National Park:

> The claim of the head of the technical department that their natural conditions 'must be preserved' is unfounded. Rather, they should be removed from the reserve and returned to the administrative authorities responsible for colonization. (Letter Baldrich to Bustillo, 3.10.1935, APN, Caja 1935)

To justify his decision, Baldrich remarkably refers to the still open debate in nature conservation, in which a distinction is made between the national park and the natural monument with regard to the protection of the natural heritage. His argument is that even the head of the technical department would have stated:

> Iguazú is not a national park, but a natural monument, and two thousand hectares would have been enough to cover this characteristic and interesting area. (Letter Baldrich to Bustillo, 3.10.1935, APN, Caja 1935)

From this he follows – against the recommendations of the technical management – "Consequently, the 50,000 hectares remaining were not necessary" (Letter Baldrich to Bustillo, 3.10.1935, APN, Caja 1935). Even though Bustillo was active in the conflict between the War Ministry and the National Park Direction for the transfer of the lands, according to the law, he largely shared Baldrich's assessment. During an on-site inspection in the National Park Iguazú, the members of the DPN Alsonso Baldrich, Luis Ortiz Basualdo, Antonio Lynch and Exequiel Bustillo probably came to the shared conclusion that "Iguazú was more than just a park, an extraordinary natural monument" (Bustillo 1997: 437).

Julio Amarante, the director of the Iguazú National Park, expresses himself in the same spirit. He wants to release 20,000 to 25,000 hectares of land for agricultural colonization. However, the future settlers must be subjected to strict moral control and discipline, so that "rigorous regulation is necessary, which should first examine the origin of the nationality of the future settlers, their personal conditions of ideological orientation and work, and which should guarantee the permanent protection of state property on the land" (Report Julio Amarante to Bustillo, 16.10. 1935, APN, Caja 1935).

Skeptical interjections, which also emphasize the value of subtropical forests, came sporadically from foresters and agricultural engineers. Thus Franco Devoto and Máximo Rothkugel point out, "Colonization and bush are considered antithetical by some technicians and practitioners" (Devoto and Rothkugel 1936: 23). But even they do not take the side of nature conservation completely, rather they recommend a model that shows different zones of use and degrees of protection within the national park:

> The solution for preserving the forest and defending colonization is to establish alternate zones, to allocate parts of the forest to permanent forest reserves, to encourage the propagation of valuable species, to sow in strips, to carry out light clearing which favors the germination and growth of valuable species and, in particular cases, to sow and plant cedar, lapacho, Petereby, laurel, etc. (Devoto and Rothkugel 1936: 27)

But such voices were too weak for sustained criticism, and so the model of agricultural colonization initially gained the upper hand. However, it ultimately failed because of its real implementation. The geopolitical concept of a population with territory voting for the instrument of agrarian colonization under state control was lacking. At the end of the 1930s, the military had to acknowledge the failure of the colonization project. In September 1941, a new presidential decree regulated the distribution of land: 500 hectares within the national park were given to private tax collectors for the purpose of agricultural colonization; the remaining areas remained in state hands under the jurisdiction of the DPN and the Ministry of War (Freitas 2018: 115-117).

In this sense, it shows how the geopolitical imaginary of the development of the "white spots" of the nation through agrarian colonization, which has its origin in the middle of the 19th century, loses its imaginary power in the 1930s of the 20th century and comes to its end. Instead, the national park discourse gained in importance, above all with the aim of developing the country for tourism. The National Park Direction was a central authority in this process – also mediated through intensive and professional public relations work – that pushed a change in the nation's imagination. From now on – according to the DPN motto "Conocer la patria es un deber / Getting to know the fatherland is a duty" – the experience of the diversity of Argentina's scenic attractions stands at the center of a new, powerful geopolitical imagination of the nation.

Cities in the Park: Urbanization at Iguazú

"National park, an area set aside by a national government for the preservation of the natural environment" defines the Encyclopedia Britannica (2019) and thus expresses an idea of national parks anchored in the common sense of most people today. In 1994, the International Union for Conservation of Nature (IUCN) carried out a categorization of nature reserves, whereby the areas protected by national parks were classified as

> large natural or near natural areas set aside to protect large-scale ecological processes, along with the complement of species and ecosystems characteristic of the area, which also provide a foundation for environmentally and culturally compatible, spiritual, scientific, educational, recreational, and visitor opportunities. (IUCN 2008: 16)

This current definition of a national park by the International Union for the Conservation of Nature contains a strict separation of nature and city, wilderness and civilization and this separation is certainly based on the U.S.-American notions of wilderness within the park movement beginning in the middle of the 19th century. Here, the national parks were seen in part explicitly as a counter-pole to the nervous stress of life in the city.

In Argentina, however, at the beginning of the 20th century, there was no striking contrast between national parks and urban development. On the contrary, state-established national parks were regarded as highlighted instruments for the urban development of peripheral regions. There are indeed voices in Argentina from the field of natural scientists, such as those of Carl Curt Hosseus and later Georges Dennler, who insisted on the establishment of classic "nature conservation parks". But these voices remained in the minority.

Even the first approaches to the development of urban infrastructure around the Iguazú Falls are linked to the development of tourism. The first colonizers, the Nuñez y Gibaja brothers, settled in 1901 after Jesuit missions had to be abandoned early in the 17th century. Also in 1901, the first organized tourist group, composed of prominent members of the patricians of Buenos Aires, came to Iguazú. A part of the group could not reach the waterfalls because of the bad infrastructure, but a member of the illustrious travel company, Victoria Aguirre, was so impressed by the landscape that she donated 3,000 pesos for the construction of a landing stage. The construction was completed with a further donation from the Nuñez y Gibaja brothers, concessionaires of the area around the waterfalls at the time. This place was then called Puerto Aguirre in honor of the noble donor, and the subsequent path towards the Iguazú Falls, a little higher up, Avenida Aguirre (Morales 1929: 95-97). Thus, the Argentine access to the today world-famous waterfalls was opened.

But then it was certainly the competition with the Brazilian neighbor that inspired the urban fantasies of Argentine bureaucrats and politicians. Brazil had already established a military colony on Iguazú in 1889, which led to a first urbanization process in this border region. The geographer Élisée Reclus describes this process as follows, "Established as a military colony, Foz de Iguazú today represents a free community, which in 1893 comprised a population of 700 people" (Reclus 1894: 390). A few years later, in December 1892, the famous intellectual Juan Ambrosetti had spoken out for the construction of Argentine military colonies in Misiones with a pamphlet (Ambrosetti 1903).

The first plan for the spatial planning of the future Iguazú National Park was drawn up in 1902 by the Belgian-French landscape architect Carlos Thays. In addition to the establishment of the National Park, the report of this expert focuses on urban planning (Freitas 2018, Berjman 2002a, Berjman and Gutiérrez 1988: 80-85). Carlos Thays started his urban planning on the already existing Avenida Aguirre. Meanwhile, he expanded the Avenida, which is now conceived as a monumental axis, into a symmetrically laid out grid bordered by a square (Freitas 2018: 108, Berjman 2002a: 30). Beyond this planning of an urban settlement, Thays also selected the suitable locations for a hotel and a casino, both of which – in the spirit of Governor Lanusse – were intended to promote the tourist development of the region.

Although the region of Puerto Aguirre has had a small settlement since 1901, which is also regarded as a relay station for visiting the Iguazú Falls, the Thays plan can be compared to the construction of a planned city because of its size. With the variation of the grid of streets, Thays uses an established planning instrument that had been taken up and adapted in different historical epochs. City historians and sociologists have also established the connection between this urban grid with processes of the colonization of space (Sennett 1991, Kaltmeier 2011: 1-19).

Specifically, Thays' grid follows an urban tradition in Latin America that can be traced back to Spanish colonial times. In the context of the laws of the Indies, the Spanish king Philipp II issued a decree in 1573, according to which all new colonial cities in Spanish America were to be built according to the *damero*, or the chessboard pattern. On the other hand, the Spanish colonial city came under increasing criticism from the 1860s onwards in the now independent Latin American nations, and the city planners more often oriented themselves to the French example of the Second Empire. Here, the monumental axes and segregated city zones used by Baron Georges-Eugène Haussmann in Paris were frequently copied elements. In this respect, Buenos Aires is regarded in Latin America as the city that resembles Paris the most. And the Avenida Aguirre could be understood as a miniature version of the Avenue des Champs-Élysées. These plans for the development of the Iguazú National Park, which was to be founded, were disseminated in the media as early as 1902, as here in a September issue of the daily newspaper *La Prensa*.

At the end of the 19th century, the planned city had a certain economic boom in the Americas. In Latin America in particular, the existing colonial urban structure proved to be too narrow for the new infrastructural and hygienic needs. In addition, rapid demographic growth – intensified by the massive immigration processes since the middle of the 19th century – set in, necessitating the construction of new cities. Ideologically, the republican elites also sought to turn away from the old colonial urban structures and inscribe themselves architecturally and urbanistically into Western European modernity, i.e. the French-influenced one.

In this context, the French architect Pedro Benoit was commissioned by the Argentine government to plan the new city of La Plata. Between 1873 and 1884, Benoit then drew up the drawing board for today's metropolis in the La Plata region. The ground plan of this planned city is characterized by a perfect square, inside which a so-called historical axis with monumental buildings is placed. The square is crossed by striking diagonals that create pyramids and diamonds accentuated by parks and squares. This general plan was awarded two gold medals at the 1889 World Exhibition in Paris and praised as the "City of the Future."

LA PRENSA – Setiembre 10 de 1902

El Iguazú

LAS NUEVAS MONEDAS FILIPINAS

RIO IGUAZÚ (Superior)

BRAZIL

First Thays Plan. *La Prensa*, 10.9.1902.

The landscape architect Carlos Thays, who was director of the Dirección de Parques y Paseos in Buenos Aires from 1891 to 1920, must have been familiar with these plans. Even if Thays' sparse personal writings did not contain any reference to this, it is not unlikely that he personally met with Benoit in the circuits of urban planning or in the French community of Buenos Aires. Central elements of the Benoit plan such as the dominant square and the monumental axis can also be found in the first Thays design for Iguazú. Only the diagonals were not used by Thays, probably this addition would have been too oversized for the planning of a small town.

Due to the unclear land ownership situation around the Iguazú Falls, however, the urban planning was not implemented. In view of the 1909 law on the colonization of the border regions, Thays had to revise the original plan and adapt it to the new conditions (Freitas 2018: 110, Berjman and Gutiérrez 1988: 80). In 1909 the plan was revised and in 1912

Thays handed over a detailed report on the *Parque Nacional Iguazú* to the Minister of Agriculture, Adolfo Mujica, richly illustrated with black and white photographs. Thays also notes the weak urban structure of the region and states:

> There is currently no resident population near the waterfalls. [...] Only in Puerto Aguirre and at the falls are there two wooden pavilions which serve as hotels. (Thays 1912 in Berjman 2002: 327)

On its renewed inspection of the region, Thays identifies a new port location seven kilometers closer to the Cataratas than Puerto Aguirre. At this location he proposes the reestablishment of the city "Iguazú". To this end, he developed the following urban planning vision:

> The total area, including the area for possible future extensions and suburbs, is approximately 900 hectares. A boulevard parallel to the river borders the construction zone, so that the beautiful river landscapes can be seen and the disadvantages of the underpopulated village near the river can be avoided. Several plazas and squares, a future hippodrome and a stadium, an airfield, etc. will be the places of recreation and movement of the new city. (Thays 1912 in Berjman 2002: 328)

In the second design of Pueblo Iguazú from 1912 – ten years after the first plans – Thays continued to pursue a geometric-symmetrical urban structure but replaced the structuring square surrounding the city with a radial design that runs towards the center of a riverside promenade and whose axis is formed by the extended Avenida Aguirre. The new plan thus contained a radical departure from the traditional, colonial Latin American urban morphology to a French-influenced neo-baroque that takes up Haussmann's building tradition. The Argentine park and urban planning experts Sonia Berjman and Ramón Gutiérrez judge accordingly:

> The Iguazú village complex is an impressive example of the fashion trends of French urban design at the beginning of the century, which attempted to combine criteria of urban aesthetics, hygiene and traffic flow by relying on the use of diagonals and a differentiated network of streets and avenues. (Berjman and Gutiérrez 1988: 83)

"Second Thays Plan". Thays 1912. Parque Reserva Iguazú.
Plano trazado general.

Perhaps Thays' new conception was also influenced by the new debates in the undisputed intellectual, cultural and political center of Argentina, the capital Buenos Aires. In 1907 and 1909, the Parisian urbanist Joseph Bouvard had visited Buenos Aires to present his Haussmann-oriented plans for the urban transformation of the Argentine capital, which also included a departure from the colonial chessboard pattern. To evaluate these plans, the city council appointed a commission of experts – including Carlos Thays. This commission came to the conclusion that it was necessary to break through the chessboard structure and instead establish "convergent and concentric roads in certain important points". Moreover, "the design of diagonal roads, a new urban panacea which would break from unbearable monotony of the old design" was necessary to give the urban ensemble a "more picturesque, varied and enjoyable appearance" (quoted by Gutiérrez 2002: 57).

While a large number of Thays' contemporaries, who were certainly dedicated to progressive thinking and who had also increasingly discovered that Iguazú Falls was a tourism magnet, had shared the ideas and plans of the Belgian-French landscape architect, there was also isolated criticism. In the intellectual field of the aristocratic elite in Buenos Aires, French cultural critic Paul Groussac expressed the sharpest criticism. Groussac had already put into question Aarón de Anchorena's presence on Isla Victoria in Nahuel Huapi Park to such an extent that he withdrew his concession and returned the island to the Argentine state. Now the second emblematic national park became the target of Groussac's sharp feather. With an undertone critical of civilization and international comparative dimensions, Groussac notes:

> Even if we look at the simple point of view of a moderately educated tourism – for this it is sufficient, for example, to consider the enormous systematic degradation of the landscape by industry or hotels in Switzerland – it is impossible, in view of this project, not to get the impression of an excess, an incongruity between the mere term "virgin jungle" and the barbarities of boulevards, squares, casinos, airports, etc., which are wasteful in the project and correspond to so many other desecrations of nature. (Groussac 1920: 233)

Overall, Groussac was highly skeptical about the tourist visions of Lanusse and Thays, as the region was too remote and climatically stressful for European and North American tourists. He also criticized the Argentine state whose task was to not create holiday resorts. However, this criticism was hardly able to gain widespread acceptance as the far-reaching urban visions in the National Park of the North failed, above all, due to the lack of state support for the implementation of the plans.

The Dirección de Parques Nacionales (DPN, National Park Direction), established in 1934 – which will be detailed later – implemented far-reaching urbanization strategies in the Patagonian south of the country. Although the urbanization efforts here – which was probably also due to the personal preferences of Exequiel Bustillo – did not reach the same central position as in the Nahuel Huapi National Park and its "capital" Bariloche. In 1935 the first national school was built in Puerto Aguirre and in 1937 the national park directorate was completed. In 1937 the extensive renovation works of the Hotel Cataratas were also completed, which included a swimming pool, tennis courts, a music salon with a grand piano, as well as vegetable fields for the supply of the hotel. In addition, the transport infrastructure was completed with the extension of the road to Puerto Canoas – the entrance gate to the waterfalls – which could now be accessed by car – and the construction of a landing strip for aircraft (Vallejos and González 2014: 22).

In 1939 the urban planning vision of Puerto Aguirre, today Puerto Iguazu, was decisively advanced. Exequiel Bustillo's brother, the respected architect Alejandro Bustillo, who had already designed the administration of the DPN in Iguazú and central buildings in the Nahuel Huapi National Park, was appointed the central planning authority. As in Bariloche, he carried out the planning in close collaboration with the architect and urbanist Ernesto Estrada, who increasingly took the lead. In its plans, DPN included central elements of the Thays Plan of 1912 for the Pueblo Iguazú at the Puerto Aguirre site, which was ultimately not realized. This includes the design of two central squares facing the river Iguazú. Along the first square, which is equipped with a viewing platform, there is a chapel, the post office, a hotel and several shops. Around the second square, facing the river, there is a school, a museum and a medical station (Berjman and Gutiérrez 1988: 87).

However, the Bustillo plan no longer had the strict consequence of the two Thays plans, but was characterized by a certain eclecticism. "From a central point of reference from which diagonals, streets and interrupted paths emanate" – at least in the diagonals one can find echoes of the radial Thays model – "a mixed set of regular and irregular blocks is structured whose extent is determined by the topography of the coast, trenches and adjacent low areas" (Berjman and Gutiérrez 1988: 91). The latter urban structure of the grid probably also coincided with the attempt to populate the urban areas. The director of the Iguazú National Park, Julio Amarante, in his 1935 report to Exequiel Bustillo, proposed a catalogue of criteria for the urbanization project, the first point of which recommended

> that the city is divided into blocks of 100x100 meters so that between them remain wide streets of at least 26 meters, including sidewalks. (Letter Julio Amarante to Director Secretario DPN Víctor Pinto, 16.10.1935, AGN, Legajo Bustillo, Caja 3344)

In it, individual urban settlers would be offered plots of land in 50x50-meter parcels.

The urban planning of the Dirección de Parques Nacionales was therefore not able to achieve the clarity and straightforwardness of the Thays designs nor that of the simultaneous Bustillo-Estrada plans in Bariloche. The same applied to the architectural vision that Bustillo and Estrada achieved in Puerto Iguazú. Although the DPN postulated a common typical colonial style in Misiones, it was unlikely to exist in this form. Instead, the architects resorted to a neo-colonial bricolage. Berjman and Gutiérrez come to the critical conclusion that:

> A 'colonial' architecture in a place where the wooden houses of the European settlers predominated, which were further developed by the Creole Republic, was a cultural intervention that was inappropriate both in terms of time and space. (1988: 92)

The DPN on site did indeed recommend preserving the architectural coherence in the planned settlements. As early as October 1935, Julio Amarante, in view of the planned urbanization, demanded, "that already on the first day the material of the building will be determined according to the standard plans selected by the Directorate of the National Parks [...] so that the interested party can choose between these models" (Letter Julio Amarante to Director Secretario DPN Víctor Pinto, 16.10.1935, APN). But the DPN could hardly enforce this demand locally.

Just as Paul Groussac had already criticized Thays' urban development plans, there were also critical voices on the architectural colonial style envisaged by DPN. Surprisingly, this criticism was expressed inside the DPN. While a certain monumentality is inherent in the neocolonial style, which should also express the colonial rule over the landscape, the foresters Franco Devoto and Máximo Rothkugel pleaded for a modest, almost humble, near-natural architecture within the Iguazú National Park:

> The building must correspond to the existing regional style and incorporate its elements. Palm leaves, boards, local stone pedestals and red clay plaster are used for small buildings. (Devoto and Rothkugel 1936: 49)

Such architecture blends harmoniously into the landscape without dominating it:

> It is necessary to propose a style, or better said, the reunification of a series of architectural elements that allow the buildings to be considered as a natural element of the landscape. [...] And this can only be achieved on the basis of 'sincerity' and naturalness in style. (Devoto and Rothkugel 1936: 49)

Simplicity and naturalness, however, were not the architectural criteria that Exequiel Bustillo had in mind for his vision of luxury tourism for the money aristocracy.

In contrast to Bariloche, Bustillo in Puerto Iguazú could not implement an overall urban development concept. Even though Puerto Iguazú was hardly able to create a harmonious, thoroughly composed cityscape, it should be noted that the landscape was reimagined in a colonial aesthetic through the construction measures. And in a way that diametrically opposed the ideas of Devoto and Rothkugel. The most visible expression of this neo-colonial resemantization of the landscape and the planning power it expresses is the building of the park administration designed by Alejandro Bustillo and completed in 1937. In this neo-colonial building, the new power of the DPN is symbolically expressed. The effect here is not only achieved through the neo-colonial architecture, but above all through the integration of the building into the park landscape. The tower of the national park administration over the river like a lighthouse reinforced by the bright white façade, and thus visible to the tourists, most of whom arrived by ship at the time.

While the DPN laid the foundations for the urbanization of the landscape from the mid-1930s, the departure of Exequiel Bustillo in 1943 and the subsequent rise of Juan Domingo Perón as Argentine President led to a renewed acceleration of the urbanization process. In 1943 Puerto Aguirre was renamed Puerto Iguazú by a presidential decree. This was followed by further expansion of the urban infrastructure, including a hospital. In addition, Puerto Iguazú's transport infrastructure was connected to the road network and in 1948 a first airport was built to replace the existing runway. However, the demographic development of the city was limited. In 1947 Puerto Iguazú had 480 inhabitants, half of them children (Berjman and Gutiérrez 1988: 91). But then it also goes demographically uphill, and so the historian Federico Freitas notes, "By 1950 Puerto Iguazú was a booming town" (2018: 121). Today, the small town of Puerto Iguazú has more than 30,000 inhabitants, most of whom live from tourism. The real development of the town at the Iguazú Falls, which was built and designed from 1901 onwards, took place only under Peronism. The connection between Peronism and Puerto Iguazú was so strong that Perón wanted to rename Puerto Iguazú after his wife, a political icon of Peronism, Evita Perón, in 1951. However, the military coup of 1955 put a stop to this renaming.

National Park Administration Iguazú. DPN. 1937.
Parque Nacional del Iguazú. Buenos Aires. DPN.

Bariloche – The Capital of the National Park

"Llao Llao is the City of the Caesars. It thus fulfills the dream of the conquerors, who had such a patriotic impression of these areas, which still had to be incorporated into the nation" (Levisman 2010: 43). With these words, the architect and architectural historian Marta Levisman sums up the significance of the luxury hotel Llao Llao, built by the National Park Direction (DPN), and refers to the inclusion of this hotel in a historical discourse of colonial origin, which presumes the existence of the legendary golden and silver city of the Caesars in the middle of Patagonia. Thus, Llao Llao is not only made a central element of a colonial and historically deeply rooted geographic imagination of the conquest of Patagonia by the architectural historian, but also an apotheosis of this narration in which myth finds its realization. Like the academic, this colonial myth has still not been critically approached by the academy in the 21st century, which makes it clear to what extent this imagination is anchored in the Argentine collective imagination.

Like no other building, the Hotel Llao Llao represents the efforts made by the Dirección de Parques Nacionales under the direction of Exequiel Bustillo to colonize the landscape through tourism. The establishment of aristocratic Gran Hotels in picturesque landscapes was in keeping with the spirit of the late 19th and early 20th centuries, as Bustillo himself noted on the occasion of the inauguration of the road network in the National Park in 1937:

> The construction of state hotels is not an innovation. Other countries have preceded us. Canada with its large Spring-Hotel in Banff, the United States with its beautiful national park hotels, Brazil with the Poço de Caldas-Hotel, Uruguay with the Carrasco and Chile with the three recently built hotels in Viña del Mar, Pucón and Puerto Varas are the expression of a practical policy that we should implement without delay. (Bustillo, 9.1.1937)

"Hotel Llao Llao before the fire". AGN, Legajo Bustillo, Caja 3354.

Llao Llao's size, grounds and orientation towards the (monetary) aristocracy of the time seem to have been clearly modelled after the Hotel Banff Springs, while the architectural design was inspired by the hotels in the U.S.-American national parks. In these parks, like Yellowstone National Park with its hotel Old Faithful Inn designed by Robert Reamer, an independent style of architectural bricolage had developed, which was later called "parkitecture" – a neologism composed of "park" and "architecture". In the USA, this architectural style was driven by the National Park Service. It was a rustic, picturesque style that adopted the rural Western European architecture, including the Swiss hut, and combined it with the architectural styles of the U.S.-American West, especially the log cabin. Andrew Jackson Dowing had introduced this Swiss style to the USA in the 1850s. He thus encountered a zeitgeist that turned to 19th-century Romanticism and took on neo-Gothic elements with nostalgic approaches. Especially in the up-and-coming money aristocracy of the Gilded Age at the *Fin de Siècle*, nostalgic European building elements that referred to the nobility were *en vogue*. Above all, they also served as a social distinction from the majority population of the popular classes (Kaltmeier 2019). This strategy of architectural distinction was found not only among the Vanderbilts, Rockefellers and Carnegies in the USA, but also among Argentine patricians such as Aarón de Anchorena. While today's unsuspecting hiker will probably be astonished when they suddenly encounter the ruins of a Roman bridge during their tours around Llao Llao, this circumstance can be explained by the park management's attempt to create nostalgic, neo-gothic worlds of experience and at the same time Europeanize the landscape.

Bustillo's comparison with Banff Springs, which had hosted such noble guests as the British Queen, does not lack a certain historical irony. Bustillo concludes his inauguration speech with the wish "we long for the same fate for Llao Llao," not knowing in what tragic way he should be right. The Banff Springs Hotel, originally built largely from local wood, burned down completely in 1926 and was rebuilt in 1928 with a new stone building in Scottish castle style. The same fate befell Llao Llao less than a year after the official inauguration on 19 November 1938. In 1939, an electrical defect led to a fire that burned the hotel to the ground, which at the time had used up the main budget of the newly established National Park Direction. About 5,000 trees had been felled and dried for the construction of the hotel (Bustillo 1988 [1982]: 42). The good networking of Bustillo and the favorable tourist situation in northern Patagonia due to the Second World War, however, led to an intensive new construction, so that Llao Llao could be completely put back into operation in 1940.

Alejandro Bustillo, brother of Exequiel, won the tender for the architectural competition to build Llao Llao's house. Exequiel's brother, however, waived his right to payment in order to cut down on speculation about favoritism and corruption. Alejandro Bustillo also drew up the plans for the reconstruction and designed 10 to 15 other buildings within the national park. In this sense, he is considered an aesthetic mentor of the architectural style used in Nahuel Huapi, which was later called "Bariloche-style" (Levisman 2010).

The Hotel Llao Llao is located at the beginning of the peninsula of the same name, between Lake Nahuel Huapi and Lake Morales, with a panoramic view over the aforementioned waters and of the majestic summit of Tronador, 25 km from the urban center of Bariloche. In short: The integration of Llao Llao into the lake and mountain landscape of the national park is breathtaking. In this respect, there were already plans in 1924 for the construction of this site (Biedma 2003: 212-213), so that the story that Alejandro Bustillo discovered this place like a rough diamond can be dismissed as another aspect of the Bustillo brothers' self-mythification (Bustillo 1997: 133-134). The hotel's panoramic complex promotes an aesthetic contemplative appropriation of the landscape, which is also based on the possibility of leisure without work and is thus closely linked to aristocratic lifestyles. At the same time, the elevated location creates the illusion of omnipotence and control that promotes colonial domination of the landscape. And it is precisely these aspects that Llao Llao is supposed to express.

The civilizing imperative of the colonization project is vividly presented in the brochure published by the DPN in 1938, which presents the progress made in public buildings. Here, in a didactic manner, the progress in the juxtaposition of "yesterday" and "today" is illustrated. Starting from the contrast between civilization and barbarity, which is deeply rooted in the Argentine imagination, DPN makes clear how the region is developing. It fundamentally sets itself apart from the indigenous, "backward" past and presents itself as highly modern and civilized with the latest infrastructure, architecture and communication media. This shows that colonization has been successfully realized through the work of DPN.

"To receive the Tourist – Yesterday and Today". DPN 1938.
Obras Públicas en los Parques Nacionales.

But the National Park Direction did not limit itself to erecting a single building. The plans for the urban development of the region went much further. What is remarkable in this sense is that the DPN even set up its own Section of Architecture and Urbanism for this purpose, headed by the architect Ernesto de Estrada in 1936. The urban planning and architectural concept, management and execution were thus the responsibility of the architect Alejandro Bustillo, as a kind of mentor, the urban planner and architect Ernesto de Estrada, who concentrated on the urban planning – including the Centro Cívico (Civic Center) – and Miguel Ángel Cesari, who supervised the construction of numerous projects (Ramos 1993). Bustillo used pencil and paper to design the first buildings in the National Park, which Cesari then completed (Berjman and Gutiérrez 1988: 20).

A main target of the urban visions of the National Park Direction was the small settlement of San Carlos de Bariloche on the edge of Lake

Nahuel Huapi, which was connected to the Argentine railway network in 1934. Bustillo clearly articulates the claim to build a new, orderly city, which acts as a bulwark of the colonization of Patagonia, and places it in a historical line that begins with the war of conquest against the Mapuche at the end of the 19th century:

> A beautiful city, well defined, with a typical regional architecture, with its amusement parks and artistic monuments. An entire human work that reflects the efforts to conquer and civilize that General Roca, Dr. Moreno, Ramos Mejía and all the great patriots of the past who turned their eyes to the vast Patagonian desert dreamed of. (Bustillo 1940 quoted from Biedma 2003: 219-220)

Now the region around today's small town of Bariloche in the middle of the 19th century was by no means a "patagonian desert," rather Mapuche cultivated wheat there, which was closely connected to the Mapuche on the Chilean side of the Cordillera via the Bariloche Pass as far as the island of Chiloé. At the end of the 19th century, the first German settlers came from Chile to Bariloche and settled there. Around 1897 – according to the German geographer Hans Steffen – there were no more than 14 settlers living around today's Bariloche. While the place itself served as a trading place for wool, established by the German-Chilean settler Carlos Wiederholdt, in exchange with the Mapuche of the region. Further settlers followed and a first pre-urban infrastructure was created. The official foundation of the city took place in 1902, which was accompanied by the state-led process of the allocation of land (Biedma 2003: 151-173). In this respect, the plans of the National Park Direction were based on an already existing small town, which in turn had neither the infrastructural nor organizational capacities for real urban planning, however DPN's urban development project certainly had the elements of a planned city.

Yet, to promote colonization through urbanization was not DPN's own idea. Bustillo was inspired by the process of the French colonization of Morocco, which was driven by Mariscal Lyautey through urban projects (Bustillo 1997: 277-279). Moreover, the good connoisseur of the history of Patagonia must have been aware of the important role of the city and fortress foundations in the Chilean part of Patagonia in the fight against the Mapuche in the middle of the 19th century (Kaltmeier 2004: 89-91).

But in Argentina's northern Patagonia, too, there had already been attempts at state-led urbanization. The most far-reaching urbanization plans around Lake Nahuel Huapi had been drawn up by the U.S. geologist Bailey Willis as part of his hydrological mission. Willis and his topographer Lewis, however, planned a completely new city – an "industrial city" (Willis 1988: 437) – at the mouth of the river Limay into Lake Nahuel Huapi. As can be seen on the map, the Limay reservoir was to be built on the north-eastern edge of the city to supply the new city with hydroelectric power. In the lower part of the city, in the direction of Nahuel Huapi, industries and the railway were to settle, while the upper part was to be marked by public buildings and shops. Here, Willis also laid a central square. Willis himself describes the planned city Nahuel Huapi as follows:

> On this level, which covers an area of 1,100 hectares, the future city of Nahuel Huapi will be built on a government-designed plan with streets, squares, parks, public buildings and water supply. This city will have sections for production, commerce and housing and will be the center of industry and transport in the Lakes Region. (Willis 1988: 212)

Willis praised the healthy climate and the outstanding natural environment of the future city, which would serve as a central gateway to the national park in addition to its industrial function:

> The location of the future city is sunny and protected, well drained and healthy. The city at the entrance to the National Park in a region with a wonderful summer climate is destined not only to be a production and trading center, but also an attractive place for tourism. (Willis 1988: 213)

This highlights that, for Bailey Willis, the establishment of the national park and the city on Nahuel Huapi were inseparably linked. Thus, the fourth chapter of his book on northern Patagonia is simply called "Lago Nahuel Huapi," and the subtitle specifies "El parque Nacional y Futuras Ciudades" (Willis 1988: 427). Geopolitically, the new city was imagined as a true "geographical center" (1988: 212), which connected the Atlantic coast with the Pacific coast and the Chilean port of Valdivia via the Andes cordillera. The city was to be a multiple development engine for northern Patagonia. Willis quotes the mission of the city to be built again from the report that he handed over to the Argentine government.

> The city will be designed according to the instructions of the Minister of Public Works, Dr. Ramos Mexía, and will be the main center of the population and manufacture of the Anden Mountains south of Mendoza. (Willis 1988: 442)

At another point Willis emphasizes the outstanding importance of Ramos Mexía for the planned colonization of northern Patagonia:

> Perhaps it was a dream, the dream of a great patriot, Ramos Mexía, but if so, it was a dream. In the beautiful setting of lake and mountain, Ramos saw a thriving city of many thousand industrious citizens, engaged in manufacturing, commerce, education, and sports, winning property for themselves and for the Republic; also receiving from their environment the vigor, the inspiration for high thinking and national enterprise. (Willis 1947: 123)

However, the plans to establish this city were not implemented, not least because the political advocate of the colonization, Ramos Mexía, had lost his influence during the political upheavals in the mid-1910s.

Nevertheless, the urbanization efforts on Nahuel Huapi 20 years later were certainly influenced by Willis, as was the founding of the park itself. However, colonization was now taking place under changed circumstances. While Willis still saw industrial development of the region that could be harmonized with the tourist park plans, the DPN was now fully committed to tourism certification. The aspects of the development of an industrial cluster that Willis envisaged no longer played a role.

But the development of the Nahuel Huapi tourism project does not go back to DPN under Exequiel Bustillo alone. In the run-up to the establishment of the Parque Nacional del Sud in 1922, in August 1918 Emilio Frey drafted a detailed, 6-page concept paper to promote tourism at the Nahuel Huapi. It contained the expansion and reconstruction of the existing rudimentary hotel structure as well as comprehensive considerations for the expansion of the transport infrastructure. The institutional core requirement was the creation of an Argentine company that organized tourism between Buenos Aires and Nahuel Huapi. Such a project was envisaged by Mr. Roth and was to receive further financial support on the basis of the concept paper drawn up by Frey (Frey, August 1918, MP, Colección Frey Bibliorato 1).

With the establishment of the Dirección de Parques Nacionales in 1934, the existing settlement of San Carlos de Bariloche became the undisputed political center of the new authority and its efforts to develop and enhance northern Patagonia through the expansion of the National Park. Bariloche had been conceived by the DPN as the administrative, economic, and tourist capital of the region, acting as a hub for political communication with the capital Buenos Aires and neighboring Chile, as well as coordinating with the smaller towns and villages in the region (Estrada [1985] in Berjman and Gutiérrez 1988: 38). This consideration was by no means new. As early as 1912, a commission set up by Ezequiel Ramos Mexía in the Ministry of Public Works began working on a study to examine the interweaving of Bariloche with the National Park of the South. In a letter to Manuel Castro, secretary of the Comisión de Fomento de Bariloche headed by Emilio Frey, the head of this commission comes to the following clear conclusion:

> There is no doubt that Bariloche will have to become the tourist center of the National Park, where hundreds of tourists will stop on their way by train from Buenos Aires over Lake Nahuel Huapi to Chile. Bariloche will thus assume the role that the important city of Lucerne currently plays in Switzerland. (Letter from the Ministry of Public Works to Manuel Castro, Secretary of the Comisión de Fomento de Bariloche, 4.4. 1914, MP, Colección Frey, Biblioratо 7)

Central elements of the subsequent discourse of the DPN, such as touristification, the comparison with Switzerland and the importance of Bariloche as a gateway to the national park, have already been expressed here.

But the merit of the DPN under Exequiel Bustillo from 1934 onwards lies in the fact that these geopolitical imaginations were now put into practice with wood, stone and asphalt. For this purpose, structural urban planning problems in the existing Bariloche had to be overcome. Like many frontier and colonization cities in the Americas, Bariloche was designed in a chessboard pattern. However, due to the steep gradients of up to 20%, there was the problem of making the streets passable. The DPN urban planning department solved this problem by constructing terraces with stairs for pedestrian traffic and newly constructed less steep ramps. In addition, a promenade was built, the extension of which was later to lead to the Hotel Llao Llao.

The masterpiece of the urban planning intervention of the DPN, however, was the complete redesign of a city center with central public buildings, which was to give the city an unmistakable face. In order to do justice to this strong administrative claim – to bring the state to a remote location on the periphery – it was the young urban planner of the DPN, Ernesto de Estrada, who proposed a complete, spatially concentrated new construction of the public buildings.

> It was necessary to combine the buildings into a harmonious whole, i.e. not to repeat the classical square of all our cities, where building plots reserved for public buildings were designated in front of the square, and to offer the rest to the public, leaving each individual the possibility of determining the type of architecture. (Estrada 1985 in Berjman and Gutiérrez 1988: 39)

The view of the lake was kept open and the new construction of this Centro Cívico could be realized by the purchase of land belonging to Primo Caparo, one of the founding fathers of Bariloche.

The architect Liliana Lolich sees the Centro Cívico in Bariloche as a prominent example of "the use of architecture to promote the state" (2007: 79). The state intervention to build a new city center is thus not a simple infrastructure measure, but a strategy to establish statehood in a peripheral region in the first place. The building itself is a prominent medium for creating the imagination of statehood. The production of statehood and the far-reaching attempt to establish national identity through architecture was pushed to the extreme by National Socialist Germany to the point where architecture functioned as propaganda. This propagandistic move, driven by the idea of colonization, can be found in the architectural and urban planning plans of the National Park Direction.

Exequiel Bustillo had enthusiastically taken up Estrada's idea of the ambitious Centro Cívico and combined it with the already smoldering idea of erecting a statue of General Julio Argentino Roca in Bariloche. Around the centennial celebrations of independence from Spain and in the face of massive social tensions caused by migration and social polarization, it was an aspiration at that time in public space – from monuments to museums – to express a homogeneous national discourse based on whiteness, Europeanization, civilization and progress (Crespo 2021). In this sense, General Julio Argentino Roca, president of Argentina from

1880 to 1886 and from 1898 to 1904, was a central figure in Bustillo's monumentalization strategy, which pursued a similar thrust. Roca was one of the main military leaders of the *Campaña del Desierto*. A war of extermination with genocidal traits waged against the Mapuche population in Patagonia. This aspect was precisely one of Bustillo's main reasons for the memorial in the Centro Cívico, the "at the same time serves as decoration and great homage to the man to whom Patagonia owes its liberation from the Indians who devastated it" (Bustillo 1997: 222). Accordingly, the central square of the Centro Cívico should also be called "Square of the Conquest of the Desert". The sculptor Emilio Sarniguet was commissioned to make the bronze statue with donations from the Jockey Club, and Alejandro Bustillo designed its pedestal. The statue was inaugurated in Bariloche on 14 January 1940. A short time later he was also to design the pedestal of the 1941 Roca statue in Buenos Aires. Only the sculptor was replaced. The motif – Roca on horseback – remained the same. However, the Bariloche statue had triggered controversy to the extent that Roca was not depicted here as a heroic warrior, but as a tired rider. José Luis Zorilla de San Martin, the artist of the second Roca statue, then corrected this representation in Buenos Aires. Today, this representation of Roca in public space by indigenous organizations and critical intellectuals, driven by the historian Osvaldo Bayer, has triggered a historical-political debate associated with a petition for a law calling for the dismantling of the Roca statue in Buenos Aires (Bayer 2010). In 2012, Mapuche activists attempted to overthrow the Roca statue on October 12, the anniversary of the conquest of America.

Mapuche activists – on the right the Chilean Mapuche journalist Pedro Cayuqueo – in front of the Roca statue and the tower of the Centro Cívico in Bariloche (Cayuqueo 2018).

While Bustillo dedicated himself to the monument of Roca, Emilio Frey did everything in his power to put a monument to his mentor Francisco P. Moreno. And so he succeeded in convincing Bustillo of this step, so that not only was a monument erected at the entrance to Bariloche, but also his remains – Moreno died in Buenos Aires in 1919 – were transferred to a mausoleum on the island of Centinela in the middle of Lake Nahuel Huapi in 1944 following a government decision in 1934. With these measures, the DPN resemantized northern Patagonia and immortalized the history of the white colonizers in the landscape through architecture and monuments.

In view of this far-reaching, still visible and highly present resemantization of landscape, it is astonishing how vague the triad of architects of the DPN Bustillo, Estrada and Cesari, all of whom were interviewed by historian Sonia Berjman, are in their answers to architectural historical questions. Ernesto de Estrada expressed himself most precisely in an interview with Sonia Berjman in 1985:

> The type of architecture has been applied on the basis of harmony with the landscape and its topography, with regional materials, rustic and taking into account the climate and colors of nature (in Berjman and Gutiérrez 1988: 41)

Alejandro Bustillo, on the other hand, portrays himself as an artist philosopher who creates something new and authentic out of himself in a quasi-genius manner. Estrada also emphasizes the importance of public institutions for civil society, such as libraries and museums. But as the photo of the inauguration of the Centro Cívico shows, civil society was not very civil. Instead, the images of the inauguration found in Bustillo's personal archive reflect a martial-militaristic component carried by the architecture of the place. This is congruent with DPN's related discourse of the colonization of Patagonia.

These images make it clear that the square of the Centro Cívico was conceived less as an open place for citizens to visit, but rather as a march-in square expressing the colonization of the region. Inside, peace and order were manifested, while – let's not forget Bustillo's fears of a Chilean annexation of the region – demonstrating his defensive strength outside.

"Inauguration Centro Cívico, 17.3.1940".
AGN, Legajo Bustillo, Caja 3354.

Such a martial vision of a strong, military state can be connected with the prevailing Germanophile proximity to the fascist regimes of the time in the conservative aristocratic circles of Argentina. Alejandro Bustillo had strongly opposed historicism and eclecticism, but especially modern architecture. Instead, he represented a sober but monumental neo-classicism, as was represented in fascist Italy or by the chief architect of Nazi Germany, Albert Speer. The Argentine architectural historian Ramos explains:

> The use of simplified classical art to express power and adopt one's own prose was part of the political project of the contemporary authoritarian states (Germany, Spain and Italy) and was justified here – by the elitist Bustillo reading – as an Argentinean way to a new level of development and splendour. (Ramos 1993)

The most visible expression of the homage to Nazi architecture is certainly the *Monumento a la bandera* in the city of Rosario, which, based on gigantic neoclassical columns and wide horizontal axes, radiates a massive monumentality that dissolves the individual in the collective. In this sense, Alejandro Bustillo's grandson recalls that he greatly appreciated Albert Speer's architecture saying, "He was very interested in this expression because he was the only one who defended the classical values born in Greece at the time, and he was an admirer of Greek culture". (Gambier 2007)

There seems to have been a certain proximity to Nazism among other people in the vicinity of Bustillo. Carlos Ortiz Basualdo, who later died in an accident in Lake Nahuel Huapi, sent an enthusiastic postcard from Nazi Germany to Exequiel Bustillo, in which he advertised German ships for Lake Nahuel Huapi, and which he ended with a "Heil Hitler!" (Postcard Ortiz to Exequiel Bustillo, 6.7.1935, AGN, Legajo Bustillo, Caja 3354).

Ernesto de Estrada was equally fascinated by the National Socialist architecture. He knew the Nazi buildings in Berlin from his own experience and had written an enthusiastic report on the Olympic Stadium in 1936 (Estrada [1936] in Gutiérrez 2007: 211-212). Estrada was particularly impressed by the "tower called the Führer" (Estrada [1936] in Gutiérrez 2007: 239), which formed the central axis in the Olympic Stadium. This alignment to a fortified tower as a central element is also found in the Centro Cívico. Here the tower was at that time the highest and most central building conceived in Bariloche and one of the central architectural landmarks of the center. In an interview conducted in 1996, Estrada describes the function of the tower as follows:

> We thought of a building with character, with a strong external image, which housed the activities of a public building inside, and we ended it with the tower as a civic symbol. In addition, this tower should attract the attention of the people and be a symbol of the place. (Estrada 1936 in Gutiérrez 2007: 126)

But obviously the Centro Cívico is not an architectural imitation of the Olympic grounds in Berlin, the Reich Party Rally grounds in Nuremberg or the planned Reich capital Germania. Instead, other influences also played a central role in the architectural design, such as the aforementioned U.S.-American Parkitecture of the National Park Service. In addition, references were made to historical city passages in Western Europe, such as the Gerechtigkeitsgasse in Bern (Lolich 2007: 69).

Overall, the Bariloche style is characterized by the eclectic use of picturesque, alpine, and rustic elements such as natural stones and woods from the region. Although the latter were used to a limited extent after the catastrophic fire at Llao Llao. Unlike detached, airy buildings such as the Hotel Llao Llao, however, the Centro Cívico has the character of a medieval fortress in Europe with its arches, corridors, tower and thick walls.

The Alpine architectural elements, which were taken up in the USA at the end of the 19th century, but which were also en vogue in Nazi Germany, influenced the style. Hitler's "second seat of government" at Obersalzberg in the Berchtesgadener Land was just as strongly influenced by these stylistic elements as the surrounding domiciles of other Nazi greats. In Nazi Germany, this architectural style – complementary to neoclassical monumentalism – was widely marketed. Hitler's "Berghof", in particular, was widely used in National Socialist propaganda (Feiber 2016: 124-139). In Bariloche, Otto Meiling, a German expatriate living in Bariloche and a convinced National Socialist at the time, called his mountain hut at Cerro Otto "Berghof" in this sense.

The National Park management, with the architectural style it shaped, flirted in this sense with the architecture practiced in Nazi Germany. In addition to use of architecture as a means of propaganda to symbolically conclude a successful colonization, the architectural heritage already prevalent in the region was certainly also relevant to the formation of the style. Coming from Chile, at the end of the 19th century mainly German colonists had settled in the region, who on the one hand brought their architecture with them and on the other were highly receptive to the picturesque Alpine style – and its links to National Socialist Germany.

Furthermore, the Bariloche style fits into a narrative that has titled northern Patagonia "Argentine Switzerland". As the historian Pedro Navarro Floria (2010) can point out, the motif of "Switzerland" was already a Patagonian myth at the end of the 19th century, which was consistently used by the National Park Direction, established in 1934, to promote the touristification of the region.

In addition, other regional stylistic devices were also used. The construction of chapels, such as La Asunción in Villa La Angostura (1936), San Eduardo in Llao Llao (1938) and Villa Catedral (1940), in particular, had been influenced by the architecture of the wooden churches on the island of Chiloé, which were declared World Heritage Sites in 2000. These wooden churches were also influenced by German, especially Bavarian, missionaries from the 17th and 18th centuries, but there is also a bricolage with Mapuche elements. At the same time the DPN recognized the close connection to the region in Chile via the Bariloche Pass established by the Mapuche.

Shortly after its official establishment, the DPN made clear its claim to leadership in Bariloche's urban planning. The 1935 Reglamento de Construcciones – a kind of land-use plan – also stipulated building regulations, such as the prohibition of non-local materials and the preferred use of natural stone. Thus, the DPN – which is not democratically anchored locally – assumed an authoritarian role in urban policy. Ultimately, the DPN was a central government authority set up in Buenos Aires that would now intervene in all planning fields in the region. Since the territory of the Nahuel Huapi National Park overlapped with that of other political administrations, such as the San Carlos de Bariloche Municipality, there were unclear responsibilities. The architect of DPN, Miguel Ángel Cesari, spoke in this respect of an agreement between DPN and Municipio, whereby DPN embellishes the city with the central public buildings and in return the city works to comply with the architectural specifications of DPN in the case of large new buildings, a pact that was later cancelled by the municipality – to Cesari's regret – and left behind, according to the architect, an urban "tutti frutti" (Cesari in Berjman and Gutiérrez 1988: 44).

DPN's urban planning aspirations extended far beyond Bariloche and covered the entire park area. While Bariloche was conceived as the capital of the region, a far-reaching concept of villages – "villas" – was conceived on a subordinate level, which opened up the park and linked to Bariloche in terms of infrastructure and administration. However, this regional planning strategy was also under the primacy of touristification, as DPN emphasizes in a brochure that states:

> The "Villas" in the National Park obey the interest of creating population centers (residential tourism) that, by serving as a point of convergence and resting on the routes, enhance the rugged beauty of the landscape. (DPN 1938)

The villages were therefore conceived as tourist points of contact, allowing tourists to explore the various sections of the National Park. A downstream goal was also the targeted settlement of Argentine settlers to help Argentineanize the border area (Gutiérrez 2007). The proposal to establish villas in El Correntoso, Lago Mascardi and Traful was already made in the DPN report from 1935. Further villas were built in Angostura, Tacul, Quila Quina, Futalaufquen and Catedral.

The DPN also undertook urban development activities in the adjacent Lanín National Park, which was founded in 1937. In 1898, General Rudecindo Roca, the brother of Argentino Roca, promoted the historic project of founding the city. This city foundation, which was connected with a former fort, was supposed to express the control over the region after the victory against the Mapuche. Until 1907, the town, which is entirely chessboard in style, was under military administration, and under the urban planning supervision of the DPN and the architectural hand of Alejandro Bustillo, it experienced a new boom. Just as Bariloche is the gateway to Nahuel Huapi Park, San Martín de los Andes is the first stop for visitors to the Lanín National Park.

Urbanization in the Nahuel Huapi National Park quickly proved to be a successful model. Between 1935 and 1940, the population grew from 1,524 to 9,800, the majority of whom settled in Bariloche, where Estrada had designated new residential areas. Today, San Carlos de Bariloche has about 150,000 inhabitants. There was hardly any contemporary criticism of the urbanization and settlement policy in the middle of the national park. Only a few conservationists, such as Hugo Salomón, opposed urbanization and, following the U.S. model of wilderness, advocated a nature reserve without human presence or use. Within the DPN, it was the forester Oscar Arce who held a similar position (Berjman and Gutiérrez 1988: 20).

National Parks and Biological Colonization

If you walk through the Nahuel Huapi National Park in early summer, you will be surprised by a true explosion of color. On the paths around Bariloche, Villa Angostura and Traful the bright yellow of the broom bushes glows, harmoniously accompanied by the rich blue of the bulging lupine umbels. Here and there, the soft pink and yellow filled pastel wine roses, which deliver the rose hips to the traditional Bariloche marmalade, give way to the full blaze of color. As idyllic as this splendor of flowers may seem to the aesthetic landscape observer, it gives ecologists and botanists a headache. For all these plants are so-called neophytes, or – as they are called by the biologists of the park administration in the National Park Direction (DPN) of Bariloche – exotic plants. The Scotch broom (Cytisus scoparius) and the wine rose (Rosa rubiginosa) come from Central Europe, while the multileaved lupin (Lupinus polyphyllus) comes from North-West America.

The spread of these new species is an expression of a biological globalization that began in European colonization. Accordingly, in invasion biology, October 12, 1492, the day on which Columbus set foot on American land, is regarded as the date for a new era of the spread of biota, based on human influences. It was only after this turn of time that neophytes were spoken of in biology, while the previously established alien plants were considered archaeophytes (Schreuder 1969: 229-230). While this classification is based on a historical-spatial dividing line, environmental historians have recently pointed to an increasing global homogenization of ecosystems. Thus Charles Mann (2011: 3-50) argues that with the biological exchange that began in 1492, the ecosystems of Eurasia, Africa and the Americas, which had been separated since the fragmentation of Pangaea by oceans insurmountable for plants and animals, are now again intertwined. In this sense, Mann understands the present era of the Anthropocene era above all as a Homogenocene. Although 1492 is rightly defined as a turning point, biological globalization also has different economic cycles. Especially in the 19th century, the internal colonization of large parts of America – such as the march to the West in the USA and the conquest of Patagonia in Argentina and Chile – led to a new acceleration of the circulation of neo-biota (for the USA see Mosena et al. 2018).

The close connection between biological globalization and colonization has been preserved to this day in the linguistic designation of the phenomenon, which is referred to as 'biological invasion', while alien biota are considered 'foreigners,' 'invaders,' 'exotics,' or 'invaders'. These terms from the military-colonial semantic environment essentially go back to the influential book *The Ecology of Invasions by Animals and Plants* by Charles Elton from 1958. This book is regarded as the founding manifesto of invasion biology and has the peculiarity that it not only describes and categorizes the phenomenon of neobiota, but also links it to further questions of conservation and the preservation of biodiversity (Davis 2006: 36-37).

In environmental history, it is especially Alfred Crosby (1972) who also starts from the turn of time in 1492 and regards this as the beginning of a "Columbian exchange" of biota. According to Crosby (2004), this exchange is an integral part of the colonial project and can be understood as "ecological imperialism". Here, the intentional exposure and settlement of alien biota is combined with its own dynamics, which these animals and plants unfold in the transformation of their new habitats. In the current debate, the use of terms with negative connotations for the phenomenon of biological globalization is highly controversial, since xenophobic terms are disseminated here and the idea of a homogeneity of supposedly autochthonous ecosystems – historically and evolutionary-biologically – that cannot be sustained is produced. But even a more neutral use of terms – such as 'alien species' – is problematic insofar as it makes the historical connection to colonial projects invisible.

Remarkably, in scientific works published before Elton's groundbreaking work, a more neutral choice of words is used, speaking of 'introduced' and 'widespread' biota or 'nonnative' plants. While Elton had the neozoons in mind, the first scientific examination of neobiota probably began in biology. In 1918, the Swiss botanist Albert Thellung introduced the concept of neophytes in a study on Montpellier. At this early stage, botanists based in Argentina were already dealing with this phenomenon on site. The first list of neophytes in Patagonia can probably be found in an essay by the German scientist Carlos Berg in the *Anales de la Sociedad científica argentina* from 1877, which identified one hundred and fifty European nonnative plants in the province of Buenos Aires and in Patagonia (Berg 1877: 183). Of the plants mentioned here,

Berg himself found 64 in Patagonia during a scientific expedition carried out between August and December 1874. Berg found a close connection between the spread of neophytes and the colonization of the south. The neophytes follow the settler and his agricultural colonization, as he points out: "Most are near the Rio Negro, where there is agriculture and horticulture, mainly wheat" (1877: 183-184). However, the phenomenon of neophytes is not yet fully understood in this essay. Thus Carl Curt Hosseus assumes that many supposed Argentinean species were introduced intentionally or unintentionally and thus have a non-native origin (Hosseus 1916: 25). He weighs up the fact that this process of biological globalization is, on the one hand, certainly for the good of the country, but, on the other hand, areas should also be protected from contact with foreign species in order to preserve the original character of the landscape.

The first fundamental contribution to a systematic discussion of the phenomenon of neophytes in Argentina was made by the Belgian botanist Lucién Hauman, who published the essay "Les Phanérogames adventices de l'Argentine" in the *Anales del Museo Nacional de Historica Nacional* in 1925. In 1926 his colleague Lorenzo Parodi published the article "Malezas de los cultivos en el partido de Pergamino" and José Molfino the article "Adiciones a la Flora fanerogámica adventicia de la Argentina", which he updated two years later with "Nuevas adiciones a la Flora fanerogámica adventicia de la Argentina".

The main work on the neophyte debate in Argentina, however, is the *Essai de géobotanique humaine* written by Lucién Hauman in 1928. "Les modifications de la flore argentine sous l'action de la civilization," which his colleague José Molfino described as:

> It is a conscientious study of human and economic geobotany with philosophical claim, in which the author, after examining the successive transformations that the native vegetation of our country has been going through for twenty years, with the analysis of relevant factors and the collection of information that concludes with a methodological list of existing adventive flowering plants. (Molfino 1928: 139)

Lucién Hauman published his essay on the *Géobotanique Humaine* as a separate comprehensive report to the Royal Academy of Belgium. As already mentioned, the subject of the 100-page essay was "Les modifications de la flore argentine sous l'action de la civilization". The starting point of his considerations is the fact that the "civilized and social man" (1928: 3) has significantly and profoundly changed the environment in which he lives in the areas of soil, water, flora and above all fauna. According to Hauman, this could also be observed in the New World and more recently in "civilized" regions such as Argentina. For Hauman, this was a glocal study – to use the neologism introduced in the spatial turn of the 1990s. For the transformations in Argentina provided information on the "general process of transforming the vegetation of our Earth" (1928: 4). In order to be able to understand the "change in the flora of a large country under the influence of civilized man" (1928: 5). Hauman sketched a comprehensive classification model of man-made transformations of flora. He categorically distinguishes between the intended and unintended introduction of neophytes. Each of these categories is then differentiated according to the geobotanical consequences. Hauman first approaches the anthropogenic interventions without value and always weighs up the contribution these actions make or the destruction they cause.

In the category of "direct actions" Hauman first lists the destruction of the upcoming vegetation without a new geobotanical contribution ("destrucction sans apports"), the result of which he understands as "exploitation". This includes, for example, deforestation – be it to exploit the wood, be it to clear the area, for example by fire. Secondly, he sees human interventions that are not accompanied by destruction but lead to the planting of new species. In this case, autochthonous plants are replaced by alien ones. Thirdly, the botanist refers to human activities that take place after the destruction of the original vegetation. Hauman understands this to mean the actual new cultivation – i.e. the agricultural and forestry use of "arable land". But – as Hauman makes clear – these direct anthropogenic interventions are only of locally limited scope. Instead, it is the indirect factors – "less visibles, but even more interesting" (1928: 9) – that have an extremely influential effect on flora change.

According to the dialectic of contribution and destruction, Hauman sees the introduction and establishment of new plants in a non-local habitat, also known as nonnative plants, initially as an emergent field of research in geobotany. On the other hand, he sees the loss of flora due to the grazing activities of domesticated animals – the controversial discussion about goats, which not only eat the photosynthetic part of the plants, but also destroy the root plexus – and therefore the phytogeography. Hauman also introduces here the phytogeographic concept of translation, which he defines as follows:

> that is, the transport of species indirectly caused by human activity; when man changes the environment chemically, physically or mechanically, he creates new habitats that colonize plants immediately through their own ways of distribution. (Hauman 1928: 9)

As the last part of this definition, which is also in italics in the original, makes clear, the plants are the actual actants here by spreading and settling. Here, Hauman assigns man a role that is only slightly reflected upon and not intended to act. He unconsciously creates an ecologically new habitat. One might think of waysides, walls, fertilized soils, or rough meadows created by grazing. The plant communities there, deeply influenced by anthropogenic use, are already referred to in contemporary botany as ruderal vegetation. According to Hauman, the lupins in Nahuel Huapi already mentioned at the beginning have also been cultivated deliberately and sometimes accidentally introduced via seeds. The lupin could then possibly spread along roadsides and colonize new areas (Hauman 1928: 26). Furthermore, Hauman mentions the targeted introduction of plants, also via *jardin d'acclimation*, as a central factor for the change in geobotany by humans. This is especially true when these plants settle in the habitats, establish themselves or even spread independently. And so Hauman summarizes and at the same time anticipates the current debate on the anthropocene, seeing man as one of the central actors in the intentional and non-intentional change of the vegetation of a region that he makes usable and colonizes (1928: 10).

Hauman sees Chile as another example of biological colonization, especially between the 37th and 42nd degree of latitude. The clearing of the Valdivian rainforest and its replacement by pasture grasses in the course of the targeted agricultural colonization from the second half of the 19th century "has remarkably 'Europeanized' certain regions of this beautiful geobotanical area" (1916: 20). Especially the German colonization of this Patagonian region from the 1850s onwards changed the landscape of the Valdivian rainforest here in just over 50 years to such an extent that "the distracted traveler could easily imagine himself driving on these roads in a Central European country" (Hauman 1916: 274). The idea of the Homogenocene formulated by Charles Mann almost one hundred years later is vividly echoed here.

Hauman's argument that it was above all the German settlers who changed the geobotany of Patagonia decisively is supported by recent geobotanical studies. At the beginning the prominent role of the wine rose in the present landscape of the Nahuel Huapi National Park was already highlighted. This is considered an invasive plant – also in other subtropical regions of settler colonialism such as Australia. Hirsch et al. were able to refute the fact that *Rosa rubiginosa* was introduced into the Nahuel Huapi National Park by Spanish colonizers via Buenos Aires on the basis of studies of the genetic material. Instead, they could prove that it was introduced around 1900 by Central European settlers. Here they suggest a direct introduction via the Italian, German, Austro-Hungarian settlers who came to Argentina between 1877 and 1897, on the one hand, and justify a probable import of the rose into Argentina via the settlers established in South Chile from the 1850s, on the other (Hirsch et al. 2011: 537).

The result remains the same. Western capitalism is making the world its own, also as far as ecosystems are concerned. Hauman sees here – entirely in the rhetoric of the 19th century that believed in progress – the civilized human being as the central actor:

> In Argentina, and probably all over the world, civilized man has settled everywhere, and since the purpose of his activity is to produce more than he physiologically needs, the result is that the civilized population, thin though it may be, quickly and deeply makes its influence felt even where half a century ago and even today there were indigenous peoples who did not cultivate the soil and only asked nature to satisfy their own animal needs, and who undoubtedly did not intervene meaningfully in its transformation. (Hauman 1928: 20-21)

Hauman speculates elsewhere that the Mapuche could possibly be responsible for the planting of apple trees in today's Rio Negro province (1928: 24), but the fundamental anthropogenic transformation takes place through the capitalist-colonizing actions of the "white man" whom Hauman calls the civilized.

Geobotanical transformation is capable of progressing to the point of complete ecocide as can be seen with the progressive settlement in the province of Entre Rios between the rivers Paraná and Uruguai, the gallery forest there – a transitional vegetation between the Mata Atlântica, the Atlantic rainforest of southern Brazil, and the Pampa – was completely destroyed by the end of the 18th century (Hauman and Castellanos 1927: 77). Against this background, botanists from the Sociedad Argentina de Ciencias Naturales were concerned with the phenomenon of neophytes and saw an opportunity to preserve the rapidly disappearing and changing vegetation geography, as well as the associated botanical diversity in the national parks or other protected areas.

Neither in Berg nor in Hauman for Patagonia are there any indications of the neophytes mentioned at the beginning of this chapter – broom and lupine. These plants – which now characterize the landscape – must therefore only have settled in Patagonia later, after the 1920s. However, this does not mean that the areas of the national parks could be regarded as a refuge for original vegetation at the beginning of the 20th century. In the Nahuel Huapi National Park, the neophytes are the red foxglove (Rubus ulmifolis), probably introduced by German settlers living in Chile, and the apple tree (Pyrus malus) (Hauman 1928: 24). Also the German-born botanist Carl Curt Hosseus identified neophytes in Nahuel Huapi and warned in his brochure on the National Park *El Proyectado Parque Nacional del Sud*, which is based on botanical results from the years 1913 and 1914:

> For us, the most important thing is to note that the vegetation of Lake Nahuel Huapí has already been completely infiltrated by foreign plants, such as the following: Madia sativa, Anthemis cotula and Potentilla anserino. It is known that on the outskirts of Buenos Aires and in the province of the same name many invasive plants have acclimatized, which has led to a change in the physiognomy of the respective floras and the destruction of the native vegetation of the country. (Hosseus 1916: 3)

The presence of alien plants was thus well known to the management of the Nahuel Huapi National Park. In a 1942 letter to the director of the park, Alexis Christensen, an employee who could not be precisely identified, argued that the U.S.-American model of wilderness, which wanted to turn the park into a "super-university of nature" in the natural scientific sense, could not be realized precisely because of the massive anthropogenic influences that had already taken place in northern Patagonia. The unknown writer makes this particularly clear in the neophytes in the national park:

> It is enough to point out that, according to botanists, 50% of the flora of our park was invaded by representatives of the European herbaceous flora, in order to consider it difficult and impossible to transform the park into a "place" of the native flora of the region. (Letter to Christensen, 2.6.1942, AGN, Legajo Bustillo, Caja 3347)

In view of this situation, it was argued that within the National Park – analogous to the National Park Act, which distinguishes between *reservas forestales* and exploitation areas – different utilization zones would need to be designated. Within the DPN, however, these use-oriented arguments were opposed by those experts who – partly on the basis of their own experience in the USA or Canada – represented the U.S.-American model of conservation. Ultimately, this was also about the very big political question of whether the DPN should be dissolved and integrated into a División de Bosques Nacionales that was more oriented towards forestry. Beyond this debate, however, the issue of biological colonization with alien species does not seem to have been particularly explosive during the peak phase of the institutionalization of national parks from the mid-1930s onwards. This may also be connected with the fact that the biologists and vegetation geographers, who had a prominent scientific interest in these debates from the mid-1910s to the 1920s, had lost their influence in the National Park Direction under the leadership of Exequiel Bustillo.

Neophytes: The Targeted Colonization of Vegetation

Today's overarching idea that national parks serve to protect autochthonous flora and fauna was also shared by natural scientists of the time but was by no means understood as common sense. On the contrary, the national parks in Argentina were considered to be privileged places where scientific research was carried out to acclimatize alien species. Scientific societies for acclimatization first emerged in France at the beginning of the 19th century and were also of great importance for the colonization of Africa and Asia, especially in England. The basic idea of acclimatization is the technical control of nature, ultimately primarily with the aim of the economic development of the colonies – but also above all in Europe of exotic aesthetic construction in gardens and zoological facilities (Osborne 2000). While the colonial acclimatization stations concentrated on North Africa – for France mainly Algeria – and for the British Empire on India as well as Australia and New Zealand, Latin America was less in the focus of European advocates of acclimatization.

However, the acclimatization movement also had an influence on the establishment of national parks in Argentina. In his 1912 report on the establishment of the Iguazú National Park, Carlos Thays already proposed the establishment of a Forestry School, a Quinta Agronómica Nacional and a zoological station, as well as experimental fields within the park area. The aim of these facilities was to research and exploit the native flora and fauna as well as to acclimatize alien species, especially sugar cane, cotton and pineapple. Thays regarded these institutions as exemplary in that the results obtained here could also be applied to provinces in northern Argentina such as Santa Fe, Entre Ríos and Tucumán that went beyond Misiones. The National Park is thus conceived here as an agricultural science and technical development cluster for the subtropical Greater Region of Argentina. In 1935, Julio Amarante, the long-serving park director in Iguazú, also advocated the planting of alien tree species for economic reasons:

> The bush is not worth much, and I believe that, as I said in my message, it should be replaced by a tall, artificial bush in which each tree legally represents a future beam. (Letter Julio Amarante to Víctor Pinto, 16.10. 1935, AGN, Legajo Bustillo, Caja 3344)

However, the acclimatization efforts in Argentina are not limited to the subtropical northwest of the country; rather, an acclimatization station was also being set up in the Patagonian national park around Lake Nahuel Huapi, or more precisely – in the middle of the lake – on the island of Victoria. The origin of this station goes back to the activities of the aristocratic landowner and adventurer Aarón de Anchorena. In 1902, he visited Isla la Victoria together with a group of excursionists, who included the later National Park Director Emilio Frey. He fell in love with this beautiful spot so much that he acquired a state concession to use it. He planted the first alien species and introduced exotic animals. In 1911, Anchorena returned the concession – including facilities such as a forest station, a hacienda and a sawmill – to the Argentine state (Koutché 1942).

Shortly after the creation of the National Park of the South in 1922, the Minister of Agriculture, Tomás Le Breton, visited the island of Victoria in 1924 and decided to set up a tree nursery there. In addition to researching acclimatization, the nursery was also intended to lay the foundations for a massive spread of suitable tree species in the region. One year later, the nursery, under the direction of Pablo Gross, began its work of planting forest and fruit trees in northern Patagonia. An undated newspaper article reports that the island was to be converted into a *reserva fiscal* and declared a national park.

In addition to praising the beauty of the landscape, the long history of the national park idea from Francisco P. Moreno to Carlos Thays and Bailey Willis to contemporary journalistic articles and the efforts of Ruiz Moreno are emphasized and placed in the context of the hemispheric national park movement with its epicenter in the USA and Canada. In the following lines, however, the economic development is dealt with – in the sense of a capitalist-colonial acclimatization logic. Once again, the close orientation towards U.S. consultants becomes clear. The U.S. ambassador to Argentina, Mr. Buchanan, had suggested the establishment of fruit plantations as well as the cultivation of hops. At the same time, reference is made to the already existing milk and cheese production with European high-performance cows and to the importance of forestry. The idea of using the island to set up an orphanage was also put forward. Fortunately, this idea, which is entirely in line with the idea of spatial separation of supposedly dangerous elements from the rest of society, was not realized (La isla Victoria. Reserva de esta tierra fiscal. Without year, MP, Colección Frey, Bibliorato 7).

After the adoption of the National Park Act of 1934 and the establishment of the Dirección Nacional de Parques Nacionales (DPN), the tree nursery passed into the hands of this institution and became part of the Nahuel Huapi National Park. In 1937, the island was divided by a fence, so that the northern part of 2,200 ha was reserved for a newly founded zoological station, while the southern part of 1,800 ha was earmarked for the Forestry Station. The latter consisted of various areas, such as the arboretum with the tree nurseries, the forest reserves with natural forests, burned areas to be reforested and pastures (Koutché 1942: 6-8). The aim of the forest station is described as "the cultivation and preparation of material for forest, rural and urban plantations in the Park's region for economic and aesthetic purposes" (1942: 9). With these criteria, it is obvious that the forest station also spread massive neophytes and contributed to biological colonization. Koutché reflects on this:

> Since the native mountain flora is not rich enough in species whose cultivation offers advantages and can thus serve different purposes, the forestry station has to search and select from a variety of foreign species those which best adapt to the regional conditions. (Koutché 1942: 9)

In a nutshell, this means:

> Studies and experiments on the acclimatization of exotic species is the second task of the station, inseparably linked to the first: plant production. (Koutché 1942: 10)

The production of the nursery is also noteworthy. As of 1 January 1941, the nursery had 474,505 plants. And in 1939 alone, 61,332 plants were sold. For 1939, the demand – consisting of sales, donations and afforestation – was 200,000 trees (1942: 22-23). Since 1925, this has been a targeted biological colonization, which has had a permanent effect on some afforestation areas and has been able to establish itself sustainably on the island of Victoria and in the region since 1937 (1942: 51).

Although there are now mainly exotic species distributed in the region, there is an awareness of the importance of autochthonous species for the pedagogy of the nation.

> In order to familiarize the visitors of the station with the most widespread and important native forest flora, the area of 5.2 hectares inside the arboretum has been reserved. (Koutché 1942: 50)

In the documents of the national park administration there is thus on the one hand a skepticism towards exotic plants, whereby the indigenous trees are preferred, and on the other hand the tree nursery that massively distributes neophytes in the region, also in the national park, without there being a detailed reflection about the phytogeographical consequences. This work does not come close to the level of reflection of the earlier work on the biological colonization by Hauman. In the 1930s, the policy of reforestation in the National Park of the South sparked an intense debate about the use of alien tree species, especially from North America. The question is simple:

> [W]ith which species should we reforest the cleared areas: native or exotic species? (Lebedeff 1932: 14)

And in his studies on reforestation in the area around Lake Nahuel Huapi, the forestry engineer Nicolás Lebedeff does not hesitate with a clear answer in favor of the native species when he writes,

> Indigenous species have great advantages, and if they are capable of satisfying the various needs of the timber market, it is necessary to give them preference. (Lebedeff 1932: 14)

In addition, the pedological fundamentals – primarily the widespread existence of volcanic ashes – favor native species. The conclusion of the forester is accordingly clear:

> When reforesting by means of artificial methods, preference should be given to indigenous species, and the exotic species should be used on a smaller scale after trials with their cultivation and in the most favourable locations. (Lebedeff 1942b: 210)

In his skepticism regarding the acclimatization of trees, Lebedeff refers to the rather limited positive experiences from Germany, where only a few species could ultimately be cultivated commercially. Accordingly, the six to seven years of experience from Isla la Victoria can only give first indications of a possible definitive introduction of new species.

Lebedeff therefore completely rejects the introduction of new species in forests that are still intact; he only considers experiments with the planting of alien species – primarily from North America – to be meaningful in completely destroyed forest corridors (1932: 16). This applies, for example, to *Pinus insignis*, which in the second half of the 20th century was to become – alongside eucalyptus from Australia – a leading plant in the forestry industry in Chile and Argentina. The good development of the test plants on the Isla Victoria is illustrated in the picture from the *Boletín Forestal* from 1942.

However, not all plantations provide forestry knowledge to the same extent. The forestry engineer criticizes the fact that the work of the Isla Victoria station so far has not met scientific standards writing:

> It is not enough just to plant so many thousands of trees and see what comes out, you have to know why. (Lebedeff 1942b: 205)

Instead, Lebedeff calls for a consistent orientation of the tree nursery and the experimental station towards the Pinchot-style U.S. forestry stating:

> The practice of forest plantations in the Nahuel Huapí National Park should be changed according to the practice in the United States. (Lebedeff 1942b: 209)

Foto Nº 15 — Diferencia en el desarrollo de Pinus silvestris (en primer plano) y Pinus insignis (segundo plano) en la misma edad de 18 años (fracción 3)

"Differences in the development of Pinus silvestris (in the foreground) and Pinus insignis (in the background) at the same age of 18 years" (Koutché 1942).

A similar skepticism towards the massive introduction of exotic trees was expressed in the mid-1930s by foresters in the Iguazú National Park in the north of the country. Franco A. Devoto and Máximo Rothkugel recommended the autochthonous plants like the black Lapacho, the Ceibo and the *palo borracho* for the beautification of the landscape around the urban facilities in the national park Iguazú, like above all the hotels. Other flowering or ornamental plants such as the yellow trumpet flower, tabebuia or the rosewood tree were not recommended as these species do not occur in the region (Devoto and Rothkugel 1936: 51). The authors have a very narrow understanding of alien plants, since all of the species mentioned were from South America and most of them from northern Argentina – just not directly in the region around the Iguazú Falls. The authors do not deal with neophytes from other regions of the world – apart from the possibility of tobacco cultivation.

If one considers the conclusions of these forestry debates of the 1930s, it is not without irony that it was precisely the forestry experts of the tree nurseries and experimental fields conceived as acclimatization stations who opposed the acclimatization of alien species and preferred native species. While nationalist motives may have also played a role here, the reasons formulated in the writings also draw on scientific explanatory patterns.

However, this scientific access also implies the possibility of changing attitudes towards exotic plants. To his great surprise, the forestry scientist Koutché, during his inspection of the plantations of the tree nursery on the Limonao peninsula in the Los Alerces National Park, found the following:

> If we now analyze the results of the sowing last spring (August-September: the normal time), we must recognize that the harvests of the native species have failed frankly and those of the exotic species are very satisfactory. (Koutché, Inspección de 1947 al Parque Nacional Los Alerces, BM-APN, Caja PN Los Alerces)

Since the soil was treated equally in both plantations and the seed was sown under the same conditions, Koutché attributes this result to the seed used, which must have been of higher quality for the exotic species. He also attributes this – with an undertone of mistrust – to the lack of experience and the lack of commitment on the part of the staff to native plants. With these explanations, the forester does not have to fundamentally revise his preference for native species.

As far as the overall classification of the forestry and agricultural stations in the national parks is concerned, however, it must also be noted that they led a niche existence, especially in comparison to the booming and protected tourism sector. In view of the fact that public internal criticism of the national park authority and its management was virtually non-existent until the departure of Bustillo, this passage by the forester Koutché must be seen – albeit cautiously formulated – as sharp criticism of the orientation of park policy:

> I have neither the intention nor the right to criticise the current policy of the National Park Direction, but I would like to point out that the forestry aspect of the Nahuel Huapí Park has been significantly delayed, especially compared to what has been done in the tourism sector. (Koutché 1942: 210)

And yet the demand for tree seedlings in the region was so great that in 1940 the DPN established another tree nursery in San Martín de los Andes in the Lanín National Park (DPN 1940: 56).

With the close institutional connection of forestry to the national park management, Argentina took a different path than its neighbor Chile, where the park movement was closely aligned with the German model of sustainable forestry (Miller Klubock 2014). The national parks there are still connected to the national forestry authority CONAF, which causes not inconsiderable conflicts of interest between the forestry industry and nature conservation within the organization. The sustainability of biological colonization and its own dynamics due to the biological invasion of the plants themselves, which is no longer regulated by man, cannot yet be clearly foreseen. The invasion biologists Daniel Simberloff, María Andrea Relva and Martín Nuñez, who in their research deal with the distribution of neobiota on the island of Victoria, which can be regarded as a kind of laboratory for biological invasions, come to the following preliminary conclusion in their study on the distribution of non-native trees writing: "The striking result on Isla Victoria is the limited extent to which this array of reportedly invasive introduced tree species has invaded the native forest" (2002: 42). But – according to the authors – these results can hardly be generalized, as in other regions of

Patagonia negative influences of foreign tree species on the native ones are evident. On the other hand, this example shows the central influence of humans and the targeted introduction of exotic plants for the change of landscapes and ecosystems.

Of Bisons and Pheasants: Landscapes for Hunting and Pleasure

Despite the skepticism about the introduction of alien tree species, there was no general national park policy on the question of the introduction or acclimatization of neobiota. For at the same time as the negative attitude of the forest experts, the national park management continued to introduce neozoon into the parks. The most ambitious project for the targeted settlement of alien species in Nahuel Huapi National Park took place in 1942. Under the direction of the National Park Direction (DPN), bison, moose and wapitis were to be transferred from the Canadian national parks to Nahuel Huapi Park. The inter-American exchange in the field of national parks was thus not limited to the circulation of ideas, knowledge, legal regulations and experts, but also extended to the planned dissemination of biota. This biological transfer was based primarily on the identification of similar ecosystem conditions, which early fueled the comparison of Canadian landscapes or the mountain ranges of the northern USA with those of Patagonia. But the main motivation on the Argentine side was probably the provision of tourist attractions, as there was no autochthonous big game beyond alpacas and pumas in Patagonia.

The selection of suitable animal species for the possible transfer to Patagonia was accompanied by Canadian experts who came to the following conclusion that "the mammals which would likely become acclimated most readily would be the Bison and Wapiti. These specimens are easily obtainable and should survive the long journey" (Letter Robertson to Santos Muñoz, 28.4.1942, AGN, Legajo Bustillo, Caja 3347).

In July the selection of the animals becomes more concrete. Exequiel Bustillo writes as follows:

> the mammals that would most likely acclimatize would be the bison and the wapiti. These specimens are readily available and should survive the long journey. (Letter Robertson to Santos Muñoz, 28.4.1942, AGN, Legajo Bustillo, Caja 3347)

In July the selection of the animals becomes more concrete. Exequiel Bustillo writes as follows:

> We would like to enrich our Zoological Station on Isla Victoria in Nahuel Huapi National Park with the following species: – two male and six female bison as well as two male and six female wapitis. (Letter Bustillo to Santos Muñoz, 2.7.1942, AGN, Legajo Bustillo, Caja 3347)

And the acclimatization station on Isla Victoria in Lake Nahuel Huapi, where the tree nursery is located, would be prepared for the exotic visitors:

> The animals will be brought to Victoria Island, where there is a zoo and acclimatization station and artificial meadows have been created for grazing and fodder production. (Letter Bustillo to Santos Muñoz, 3.3.1942, AGN, Legajo Bustillo, Caja 3347)

This zoological station was established in the mid-1930s under the direction of the National Park management, primarily for the breeding of native species such as the small deer species Huemul and Pudú, maras, pumas and condors, but was also intended to serve as a tourist attraction (Municipio de San Carlos de Bariloche, without year, DPN 1936: 11-12). And so the National Park Authority promoted the work of the Zoological Station as well as the rearing of alien red deer in a self-portrait brochure from 1938. The first colony of 109 deer was donated to the park by Aarón de Anchorena. Their transport to the park and acclimatization took place without any problems (DPN 1936: 12).

In the Lanín National Park, exotic animals were also introduced in a targeted but sometimes unintentional manner, as already described in detail in a 1980 report on the National Park. Thus the first deer (Ciervo elaphus), which came from Austria-Hungary, were introduced in 1906 by the landowner Pedro Luro in the province of La Plata and their descendants were brought to his hacienda near Junín de los Andes by Roberto Hohmann in 1922. From there they spread – largely without natural enemies – to the province of Neuquén and thus also to the later Lanín National Park. The main reason for the introduction was hunting, as antlers are considered a coveted hunting trophy. The European wild boar was also brought to Argentina by Pedro Luro to populate the hunts of his hacienda. From there – according to the hypothesis of the unknown author of the National Park brochure – the adaptable wild boars spread as far as Neuquén (El Parque Nacional Lanín 1980: 35-36, BM-APN, Caja PN Lanín).

"The Zoological Station". DPN 1938. *Obra Pública, cultural y turística realizada en los Parques Nacionales*. Buenos Aires.

The introduction of alien species was based on an idea that saw Nahuel Huapi Park as an empty space to be colonized. On the one hand, reference was made to the previous extinction of autochthonous fauna in the national park:

> As for Nahuel Huapí, it is important to emphasize the fact that native species have almost completely disappeared due to a number of factors that cannot be analyzed here. (DPN 1936: 11)

Because of this, native species were also reared in the Zoological Station. On the other hand, however, the autochthonous fauna was also described as deficient and species-poor. In the eyes of the park managers, this justified the targeted introduction of neozoon. Originally, DPN's acclimatization efforts were limited to the island of Victoria. The comparison to Jurassic World Island is obvious, also with regard to the safety against eruption. As early as 1938 the DPN stated:

> The deer released during 1936 have become acclimatized and their presence on the island has been recorded in group sizes that vary in their exact composition, but groups of ten or twelve individuals are abundant.

> Recently, these animals have been seen in cattle farm No. 26, so it can be concluded that they swim across the lake, possibly on hot days, so it is not surprising that over time the park has become populated by this species. (DPN 1938: 121-122)

The distribution of European red deer and their browsing eating habits has had a clearly negative influence on tree growth. In a four-year long-term study, the above-mentioned inter-American invasive biological research team Nuñez, Relva and Simberloff recently demonstrated that both the native tree species and the exotic ones on the island of Victoria are negatively influenced in height growth by game browsing. The native Chile cedar (Austrocedrus chilensis) has a much stronger negative influence than the exotic Douglas fir (Pseudotsuga menziesii). The scientists interpret this as an indication that could confirm the "invasive meltdown hypothesis". The rationale behind this hypothesis is that non-native species support each other in the biological invasion and thus mutually increase their probability of survival by changing the ecosystems in their favor. In this sense, the reverse conclusion has also been drawn that the targeted removal of European red deer and the reintroduction of native species such as Pudú and Huemul strengthen native plant communities (Nuñez, Relva and Simberloff 2010).

While the beautiful new world of biological homogenization has become a thorn in the side of today's nature conservationists, the increase in attractiveness through neozoon aimed at by the park makers was already bearing fruit in the 1930s and 1940s. The bricolage of deer, exotic-looking Arrayanes forests and a West European-style log cabin also created another tale about Nahuel Huapi Park. Invited by Antonio Lynch, aristocratic member of the park commission, the U.S.-American director Walt Disney travelled to Argentina in 1942 to stay with Lynch in his simple log house in this forest. Lynch was the owner of Quetrihué, a peninsula in the middle of Lake Nahuel Huapi, characterized by one of the last surviving Arrayanes forests, which in 1972 became Arrayanes National Park in the middle of the Nahuel Huapi National Park, a national park in a national park, to preserve its special characteristic trees. The Arrayán tree, which grows to a height of 15 meters, is most recognizable for its cinnamon to brick-red bark, which gives the forest a fairy-tale-like character. The walks and hikes among these trees inspired Disney to create the iconic blockbuster "Bambi".

However, returning to the story of the Canadian moose, the new North American animal inhabitants of the South American Zoological Station were to come from the Elk Island National Park in the Canadian province of Alberta and were probably given to the Argentine government free of charge. However, the government would have to bear the shipping costs. Due to the climatic conditions, a shipment was planned for September (Letter Robertson to Santos Muñoz, 28.4.1942, AGN, Legajo Bustillo, Caja 3347). In October, Exequiel Bustillo also renewed Argentinean interest writing, "Elks or any other species of deer other than wapiti, the introduction of which into the country is prohibited by the General Direction for Livestock" (Letter Bustillo to Santos Muñoz, 9.4.1942, AGN, Legajo Bustillo, Caja 3347; 10.1942, AGN, Legajo Bustillo, Caja 3347) to the Nahuel Huapi National Park. While Bustillo expressed himself very enthusiastically in national park questions, he did not seem to be very enthusiastic about the transfer of the animals. Instead, in the letter to the Argentine ambassador in Canada, Dr. Pablo Santos Muñoz, he repeatedly refers to Aarón de Anchorena as the actual initiator of the naturalization of the North American big game in Patagonia. Anchorena was known in Argentina for the acclimatization of new animal species, which corresponded to an aristocratic lifestyle of the time. In his residence "Estancia Anchorena" on the banks of the Rio de la Plata in Uruguay and the Rio San Juan – declared Parque Nacional Anchorena in 1978 – Anchorena introduced the Indian axis deer, Australian kangaroos, and Northamerican buffalos as well as European hunting game such as wild boars, deer and pheasants at the beginning of the 20th century. Anchorena, who had been using the island of Victoria since 1904, had also introduced these exotic species to Patagonia, along with farm animals such as sheep, Dutch cattle and horses. Anchorena was forced to return the rights to use the island to the Ministry of Agriculture in 1916 after being attacked by Paul Groussac. However, the animals remained. Today, the red deer, the wild boar and the European hare are regarded as alien species established in the Nahuel Huapi National Park.

The aim of the biological colonization of the Nahuel Huapi Park under the Bustillo-led DPN was most likely the ultimate transformation of the wild, indigenous landscape into an aristocratic landscape of European – and also specifically English – character. In these landscapes, tourists should be able to experience a spectacular animal world and pursue the aristocratic adventure of hunting. Another element of the aristocratic Europeanization of the landscape is the introduction of highly valued bird species such as pheasants for hunting and swans for aesthetic edification. Fifty swans were abandoned in Lake Nahuel Huapi as part of a donation of the Buenos Aires Zoological Garden made by Exequiel Bustillo in 1935 (DPN 1936: 123).

In Iguazú National Park, the tourist valorization of fauna also took shape. In the subtropical rainforest landscape, however, the park administration relied on exoticism. Thus, in 1939, a Zoologic Park was established in the national park, which probably had no scientific claim, but was only supposed to display animals. For example, the DPN did not make any effort to ensure taxonomic accuracy in its inventory, and simply lists one monkey and two turtles in its stock list, without naming the exact species (DPN 1940: 34).

Biological Colonization of Patagonian Waters

Yet not only the land was biologically colonized. The waters of Patagonia in particular were heavily populated by new species in the context of the development of peripheral areas by national parks. At the turning of the 19th to the 20th century, fish culture in inland waters was one of the most striking transnational projects of biological colonization. In the first years of the 20th century, experiments on the acclimatization of edible fish, especially from the salmon family, took place in the austral regions of the world – from Chile to Argentina to New Zealand. In this large-scale experiment, fish spawn from North America and Western Europe – the northern hemisphere – were introduced into the waters of the southern half of the world. The Patagonian region of Argentina played a pioneering role here.

In Patagonia, the introduction of alien fish species to Patagonia was initiated by Francisco P. Moreno. The scientist and discoverer thus proved to be visionary for the future development of the region, as he had already done with regard to the national park projects. In 1892, Moreno invited the French scientist Dr. Fernando Lahille to Argentina to record the ichthyological fauna of the region and to prepare the acclimatization of alien species (Marini 1936: 9). Lahille later became head of the Museum de La Plata and head of the Department of Hunting and Fish Farming of the Argentine Ministry of Agriculture.

The peak of the biological colonization of Patagonian waters by edible fish species from the northern hemisphere was in the first decade of the 20th century. It was characterized by a high level of commitment on the part of state authorities, which proactively sought cooperation with U.S. aquaculture experts. In 1900, the Minister of Agriculture, Dr. M. García Merou, promoted the acclimatization of fish and commissioned the Italian natural scientist Felipe Silvestri to disseminate new fish species in Argentina. On the basis of these studies, the idea of introducing alien fish species, especially salmonids and trout, was decisively continued in 1903. García Merou, who then held the post of Argentine ambassador to Washington, made contact with the U.S. authorities and commissioned John W. Titcomb, at the time head of the Department of Fish Culture in the United States Bureau of Fisheries, to bring U.S. salmon species to northern Patagonia (Marini 1936: 10). He examined the waters of northern Patagonia and selected the appropriate alien fish species to colonize the Patagonian waters.

But then the fish culture experts faced another challenge: the transport of the fish spawn. The direct transport route from the U.S. east coast to Buenos Aires was not possible because the ships heading directly for Buenos Aires did not have the necessary cooling chambers. Instead, ships equipped with cooling chambers used for bringing Argentine beef to England were used because of their ability to cool the delicate spawn to a constant 4 °C. According to Marini, who in the 1930s headed the Department of Fish Farming at the Argentine Ministry of Agriculture, the shipment of fish spawn, organized in 1903, was the first to take place over such a long distance. The journey of the fish embryos lasted 50 days, whereby altogether in 30 days about 16,600 km had to be covered by sea and a further arduous 1,600 km by land from Buenos Aires to the Nahuel Huapi lake. The journey of the fish spawn is thus quite close to the strains of Phileas Fogg and his servant Passepartout, who crossed the world in 80 days in Jules Verne's famous novel.

With meticulous planning, the success of this action was overwhelming. Accompanied by the U.S. expert E.A. Tulian, who had been specially contracted for this purpose, seven container boxes of fish spawn from the USA arrived in Patagonia in 1904. The loss rate of 10% of the spawn – especially in view of the exhausting journey – was extremely low. However, the second transport of fish spawn under the leadership of U.S. expert Ormsby in 1904 highlights how fragile the cargo is: 90% of the spawn was lost in this shipment because the fish hatched off at the Brazilian coast prematurely (Tulian 1910).

In Patagonia, the Argentine government set up a fish breeding station at the Rio Limay, which received the fish spawn. Initially headed by U.S. experts, the central figures here were first John W. Titcomb and then E.A. Tulian, who were contracted by the Argentine government to head the Department of Fish Culture at the Department of Agriculture. Tulian held this post until June 1909. As a pioneer of aquaculture in Patagonia, Titcomb chose a site with an underground spring at the Rio Limay, which regularly produced cool fresh water, near Lake Nahuel Huapi. He also took over the first management of the station but returned to the USA the following year to organize further transports. Instead of him, other U.S. experts took over the management of the station in Bariloche. At first this was Mr. Ormsby; then Mr. Brosby followed, who was to lead the breeding department in Patagonia for almost 20 years. Thus, the development of aquaculture in Patagonia is closely linked to the work of U.S. experts who prepared the terrain, selected the fish, organized the transport and acclimatized the fish.

In 1909, under the direction of L. Valette, the first transport of fish for spawning could take place without the assistance of U.S. experts. Valette then took over the management of the Department of Fish Culture in the Ministry of Agriculture after Tulian's resignation. But the freshly trained Argentinean fish breeders did not want to do without the US-American expertise. And so, the fish breeder H. Kelly was hired, among other things to supervise another delivery of fish spawn (Marini 1936: 17).

But let us now turn our attention to the actors of the biological invasion, the fish. Specifically, the first delicate cargo in 1903 consisted of 1,000,000 larvae of herring vendace (Coregonus clupeaformis) and several thousand larvae of brook trout (Salvelinus fontinalis), American lake trout (Salvelinus namaycush), and Atlantic salmon (Salmo salar) and its inland variant (Salmo solar m. sebago). Herring whitefish was only released in Lake Nahuel Huapi, the others in Traful, Espejo and Gutiérrez lakes (Marini 1936: 11), the latter two being connected to the Nahuel Huapi by rivers. In the following years until 1910, in the context of the new fish farming department and in cooperation with U.S. experts, further larvae were reared as young fish and introduced into Lake Nahuel Huapi.

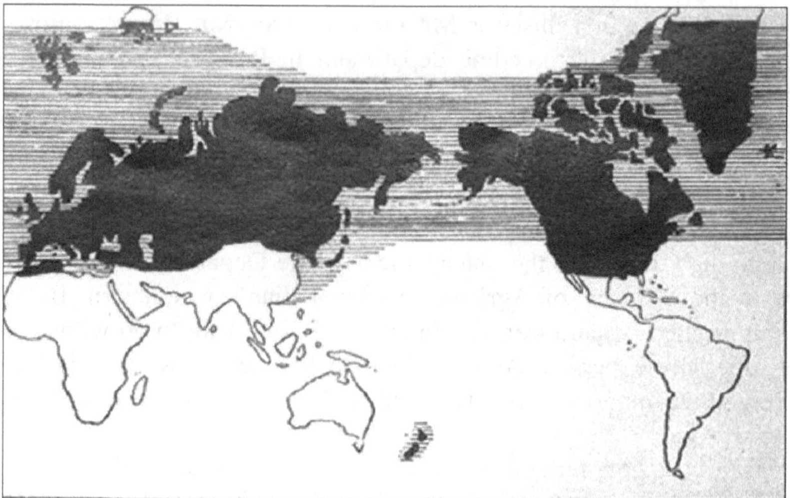

Fig. 1. — Planisferio que demuestra el área de dispersión de los salmónidos.

"Global Distribution of Salmonids" (Marini 1936: 19).

Fig. 2. — Forma en que fueron transportados los primeros embriones de salmónidos introducidos desde Neuquen hasta San Carlos de Bariloche, penosa travesía que duraba alrededor de 20 días.

"Transport of fish spawn from Neuquén to San Carlos de Bariloche" (Marini 1936: 19).

Because the main aim was to introduce new species, Tulian was commissioned several times in Europe – namely in England, France, Belgium and Germany – to acquire new fish species. The main focus here was on rainbow trout, the introduction of which had proved problematic during previous transports, and the European brown trout (Salmo trutta fario). Estimates predicted that by 1930 about 9 million eggs or larvae of alien species – especially salmon-like ones – would be introduced into Argentine waters (Reed 1949: 3-4). In a 1908 lecture, Tulian reports the following successful results of acclimatization:

> Lake trout were found in Lago Traful and Lago Correntosa, and land salmon in Lago Gutierrez, while brown trout were found in almost all rivers and streams. In many of them the brown trout are very numerous and increasing rapidly. The inspector and assistant of the fish farm Nahuel Huapi took, both last year and this year, thousands of fingerlings from irrigation ditches that draw their water from these streams and put them back into the streams. 860 fingerlings of brown trout were taken only in April last year from an irrigation ditch of a small garden that is on the way to the "arroyo de Jones" and 2,300 from another one that flows into the "arroyo de Newbery". These were undoubtedly fingerlings that hatched in September or October 1907. (Tulian 1910: 958)

This successful acclimatization of North American salmonids was also noted by Bailey Willis during his hydrological mission to Patagonia. The U.S. geologist writes enthusiastically, "The fish have multiplied so much that they are now abundant and have reached an extraordinary size" (Willis 1988: 290). The size and quality of the fish is – as the fish farming experts also emphasized – highly satisfactory. Especially the Traful lake produced outstanding fish – in fact employees of the fish breeding station caught a salmon that weighed 15 ½ kilograms (Reed 1949: 8-9).

Concern still existed for the salmon as to how the migration into the oceans and the possible return would look like. Experiences from New Zealand, where salmon were acclimatized almost simultaneously to Argentina, show, however, that the introduced animals migrated back to their source inland waters (Marini 1936: 15). Once again, the transnational dimension of the contemporary practice of aquaculture is revealed here.

Argentine fish farming soon became independent of spawning imports from North America and Western Europe. In 1907 the fish breeding station at Lake Nahuel Huapi produced its first own offspring. Marini proudly reports:

> It was the first time in South America that a whole school of brown trout had been spawned, producing 270,000 eggs which developed under perfect conditions. (Marini 1936: 12)

For the 1908/1909 season, Tulian assumes that the Nahuel Huapi fish farm will be able to harvest half a million eggs of brook trout spawn and several thousand salmon eggs (Tulian 1910: 959-960). The success of Argentine fish farming spread rapidly throughout the region. And so, the first Chilean fish breeding station on the Rio Blanco, established by Federico Albert in 1905, ordered fish spawning from the Nahuel Huapi region as early as 1908.

Current invasive biological work clearly shows that 8 of the 10 salmonids introduced in the Argentinean part of Patagonia can be considered established. However, strong regional differences and historical variations must also be taken into account. At first, the brook trout was regarded as a particularly invasive species, but it was increasingly replaced by the rainbow trout (Macchi and Vigliano 2014).

With the targeted biological colonization of the waters, Marini, and thus also the national park management, were fully aware that they were massively altering the ecosystems in question. In this sense, Marini formulates *avant la lettre*, the idea of the Homogenocene, developed by Charles Mann writing, "The faunas, both terrestrial and aquatic, will converge over time" (1936: 7). However, Marini plays down the danger for the domestic four to five fish species saying, "It turns out that it is not alarming that some species of salmonids have been introduced into this environment, as the number of native fish have always been very limited" (1936: 8). But later studies show that the native Puyen (Galaxia maculatus) is a preferred food source for the introduced salmonids stating:

> It is very common to observe in the stomach contents of salmonids an extraordinary amount of Puyen, which forms a true mass that occupies the whole stomach. (González Regalado 1945: 126)

At the same time, however, Marini clearly has a latent awareness of the threat that the fish species used pose to the local population. Marini makes the following distinctions accordingly:

> I believe that, when new fish imports are introduced, white fish should be taken very seriously, because they are small-mouthed salmonids that cannot eat other fish and their acclimation in places where there is the Pejerrey prevents the development of indigenous species from being affected. (Marini 1936: 12)

This seems to be confirmed by recent studies which attribute the successful invasion of rainbow trout precisely to their ability as predators feeding on the native Galaxias maculatus (Macchi, Pascual and Vigliano 2007). This gives them an advantage over other exotic species, such as the brook trout, which after a first massive introduction declined most likely due to competitive pressure in the population, as well as the native species (Percichtys trucha) and Odontesthes hatcheri. As recently as 1945, the biologist Tomás González Regalado noted a very wide distribution of both native species in the Patagonian national parks Nahuel Huapi, Lanín and Los Alerces (González Regalado 1945: 121-126). However, due to competition in the search for food and the predatory feeding of invasive alien salmonids, the populations declined massively. State efforts to establish aquaculture in Patagonia with alien fish were driven by the idea of economic development of the remote region. The first efforts were successful in that the alien species successfully colonized the Patagonian waters. But apparently the further economic development suffered from the lack of market access, which can be attributed to missing infrastructure, the lack of regional demand due to the low population density in the region and also to the eating habits established in Argentina.

In this sense, the still young Argentinean fish culture was in a stalemate in the 1930s. In view of the massive efforts made by the DNP under the leadership of Bustillo to develop tourism in northern Patagonia, the conditions for fish culture also changed. Around Lake Nahuel Huapi, Bustillo created a touristic landscape that corresponded to the needs of money aristocratic holidaymakers.

This also included the promotion of the landscape as a paradise for anglers. In the 1930s, this vision was so successful that even the head of the Department of Fish Culture in the Ministry of Agriculture, Tomás Marini, could not escape this geopolitical imagination. During a conference organised by the Sociedad Científica on 8 November 1935, he stressed the importance of the aquatic neozoon for tourism:

> The salmon fishing in our Parque Nacional del Sud is one of the changes that will surprise the beginner in this sport and satisfy the wishes of the most experienced. In any case, it will arouse such enthusiasm that tourists will unfailingly strive to return to places where a sport can be practiced that evokes such intimate emotions that can only be explained if you have experienced it yourself. For this reason, salmon fishing is and will always be one of the major attractions that Nahuel Huapi Park offers tourists from the point of view of anglers. (Marini 1936: 2)

With these words, Marini links the introduction of alien fish species via the tourism node with the establishment of the National Park. Here it becomes clear that the acclimatization of exotic species is booming not despite but precisely because of the National Park. Especially in the field of fish culture, the National Park is proving to be a neozoon catapult.

But Marini also makes another important point in his speech, which expresses a swelling controversy. For Marini, sport fishing is only a side aspect of the use of fish wealth. At the same time, as a representative of the Ministry of Agriculture, he advocates the continued commercial use of the imported salmonids. This economic development starts with the supply of the surrounding hotels and the local population, and then leads to far-reaching plans for the export of fish. The aim is to fully exploit the economic potential of fish farming (1936: 8). For this purpose, however, the other waters in northern Patagonia must also be used.

> All these waters, and many more that should be long listed, are waiting impatiently, as I said, for our work, to generously give us back one hundred fish for a single fish brought in. And there is no doubt that in all these waters, in addition to the lakes of the National Park, the various species of salmonids will become extremely important. (Marini 1936: 4)

However, the park management relied solely on sport fishing and, in the Reglamento de Pesca del Parque Nacional de Nahuel Huapi of 28 July 1935, prohibited fishing with nets or dynamite as well as all commercial fishing (DPN 1936: 103, Bustillo 1997: 349). For the elaboration of this decree Bustillo contacted a colleague from the USA who sent him the hunting and fishing decree of the State of New York.

Obviously, the fish farm of the Ministry of Agriculture did not agree with this regulation. In 1939, a ministerial decree issued by the fish farm in the Nahuel Huapi National Park allowed the sale of fish to hotels in the same area. This was intended to partially cover the costs of the station. At this point, a long-lasting conflict broke out between Marini, head of the Department of Fish Culture in the Ministry of Agriculture, and Exequiel Bustillo, head of DPN. Bustillo accused the fish farming station of commercialization leading to the destruction of the "ichthylogical heritage". In 1942, in a letter to the Minister of Agriculture, he vehemently spoke out against this regulation. According to Bustillo, the commercial use "seriously damages the fish stocks of the lake district, which not only limits the possibilities for sport fishing, and which is one of the main attractions for tourism" (Letter Bustillo and Antonio Lynch to Dr. Daniel Amadeo y Videla, Minister of Agriculture, 4.3.1942, AGN, Legajo Bustillo, Caja 3347). And he was not afraid to publicize this debate through the press. In defense of his position, Bustillo cited the famous angler Sir Neville Henderson, ambassador of Great Britain to Argentina and later to Germany (Bustillo 1997: 353-355).

This conflict also has an institutional dimension. In this specific case, there was a conflict of competence over the management of the fish culture in the Nahuel Huapi National Park. Bustillo was aiming to integrate the Fish Farming Station into the National Park Direction, while Marini considered it an integral part of the Fish Farming Department of the Ministry of Agriculture. (Bustillo 1997: 350-352) Ultimately, Bustillo seems to be intent on establishing DPN as the only relevant state authority in the territories of the national parks. In the end, Marini was hardly happy with the relocation of the first, rudimentary fish rearing station on the Rio Limay to the newly built and larger "Piscicultura Bariloche" in 1932, as there were less favorable water conditions here. And – it can be assumed – the proximity to DPN, which was striving for dominance, was also more pronounced. In 1971, this station was transformed into the "Centro de Salmonicultura Bariloche," which was finally closed in 1994 (Macchi and Vigiliano 2014: 167).

For Bustillo, the promotion of sport fishing was a central element in the overarching strategy of the European aristocratic colonization of the landscape through tourism. The importance and status of fishing in the circles of the cosmopolitan (monetary) aristocracy should not be underestimated (Kinsey 2008: 17-27). Anglophone gentlemen from the nobility or the military elite in particular devoted themselves to this genre and spread it across the Americas so that the National Park Authority also published its own English-language fishing guide in the 1940s.).

In its basic structure, this conflict between aquaculture and sport fishing resembles the smouldering conflict between forestry and the park landscape. While technicians from the context of the Ministry of Agriculture focused on the economic development of natural resources, for Exequiel Bustillo the development of the region through tourism was the most important. Genuine nature conservation issues were, however, of only marginal importance for both positions.

"Fishing at Nahuel Huapi" (Reed 1949: cover picture).

"What is a National Park?"

The U.S.-American geologist Bailey Willis asks himself "What is a national park?" in a 1913 brochure in which he designs a comprehensive program for the establishment of the National Park of the South in Argentina. To answer this question, Willis draws on the scientific instrument of definition. And he notes:

> A national park is an area reserved by the state for the leisure and well-being of the entire population. It is set up in such a way that no single person can prevent others from enjoying it and preserving everything that corresponds to human practice in its natural state. (Willis 1913)

Willis is not only concerned with capturing the facts in a way that can be understood by others, the definition also aims to determine the nature of the national park universally – independent of time and space. For the question posed in this book of the location of national parks in a transnational interweaving space and the orientation of parks in the field of tension between nature conservation and colonization, the determination of a universal, true essence of national parks proves to be too limited. Therefore, I do not want to ask about the inner essence of a national park, but to ask the question differently: How does the National Park operate?

With this changed question, a whole arsenal of discourses and concepts, practices and regulations, effects and materializations are brought into focus. These aspects – from the transnational circulation of environmental knowledge, scientific experts, tourists and neobiota, to biopolitical practices of regulating the relationship between population and territory in the park, to materializations such as the construction of planned cities and huts for park rangers – have been extensively analyzed and discussed in this book. In addition, this book has explored how the interdependence of individual practices and discourses leads to the production of overarching, in part heterogeneous geopolitical imaginaries, which in turn circulate via media dissemination in the transnational space of interdependence, such as the imaginarium of Patagonia as Switzerland of Argentina or that of the civilized aristocratic park landscape.

Nevertheless, the question arises as to whether the functionality of the national park in Argentina between 1903 and 1943 could not be narrowed even further. In short, my aim here is to find an apt conceptualization that has neither the ontological narrowness of a universalist definition of essence, nor the heuristic openness of a game of various practices, discourses, and materializations.

From Foucault's perspective, the national park could possibly be understood as a dispositive. Foucault understands the dispositive as:

> a decidedly heterogeneous ensemble that encompasses discourses, institutions, architectural institutions, regulatory decisions, laws, administrative measures, scientific statements, philosophical, moral or philanthropic theorems, in short: both spoken and unsaid. As far as the elements of the dispositive are concerned, the dispositive is the net that is woven between these elements. (Foucault 1978: 119f.)

In this sense, the National Park can certainly be considered as a dispositive, since the analysis to date has taken into account precisely these elements named by Foucault. In addition, Foucault links the emergence of dispositives with crisis situations, which are also present in the case of the Argentine national parks with regard to an incipient ecological crisis and above all with regard to the crisis of governance of the peripheral regions. However, Foucault's dispositive is conceived as a power ensemble that permeates all areas of society and thus thinks of power decentrally and not from a central authority such as the state.

The national park, on the other hand, can hardly be imagined without the nation state. This was recently underlined in the environmental historical debate with the concept of the "nature state" (Hardenberg, Kelly, Leal and Wakild 2017). Despite the importance of statehood, however, caution is also required to reproduce the imaginary of a coherent state that permeates all areas of society, produced by the state itself, and thus to sit up with the state as a fetish. In this respect, the National Park cannot be conceptualized as a state apparatus due to its far-reaching interdependence with civil society, other communities and academia, and its inherent logic.

A special feature of national parks is their specific territoriality. They are characterized by a spatially sharp difference between inside and outside, marked by the boundaries of the park. Thus it would be tempting to understand the national parks as heterotopic enclaves. But even such an enclaved conceptualization would hardly do justice to the national parks, since it would neglect the transnational circulation of knowledge, objects and biota that has just been worked out in this book.

In view of the unsatisfactory attempts to conceptualize Argentine national parks in the first half of the 20th century, I would now like to propose that the national park be seen as a bridgehead. The concept of a bridgehead goes back to the dependency-theoretical discussion of the 1970s. It reacts to a too simple, dualistic juxtaposition of center, or metropolis, and periphery and thus justifies the structural heterogeneity in the Global South. The bridgehead argues that the interests of the center are also represented within the periphery. This refers above all to the social elite in the countries of the periphery, which does not represent the interests of the majority of the population there, but the interests of the center. This is not the place for a detailed critical discussion of this concept. Instead, I would like to address four aspects that clearly show how this concept can be used fruitfully for understanding the functioning of Argentine national parks.

First, there is the tendency towards colonization, which is also laid down in the dependency-theoretical determination of the bridgehead. The concept of the bridgehead is borrowed from military language and designates a kind of vanguard or docking point for further military actions that serve expansion and conquest. This proactive element of the conquest of previously uncontrolled areas was already established in the national parks in the north and south sketched in 1902 and 1903, but was then brought to a point by Exequiel Bustillo, following the work of Bailey Willis, from 1934 onwards, when the national park became a "true instrument of colonization". In this respect, the bridgehead is not static, but rather characterized by the tendency of expansion, colonization and land seizure.

Secondly, there is a power gap in the bridgehead between the center and the periphery. In this way, the close connection between the national park and the state can be conceptualized. However, the state is by no means to be understood as a fixed object or even subject. Rather, it is based on a dynamic understanding of the state according to which the various state apparatuses are themselves constantly fought over and negotiated, so that the state itself can be understood more as an effect of the various struggles and government practices. In this sense, the state also does not control the claimed national territory on a comprehensive basis. Rather, it must make itself visible and put itself in the public eye. This applies all the more to the peripheral areas of the state in which nature reserves are located. Based on these considerations, the national park can be understood as an instrument that creates statehood in the peripheral areas first. The imagination of the nation state is thus produced via the various techniques of regulation, demarcation, statistics and image circulation in the national parks.

Third, the bridgehead concept implies a heterogeneous spatial dimension. This is relevant in two respects. On the one hand, it is an extension of the center that is spatially interrupted but maintained by communication channels. The latter ensure the permanent circulation of ideas, people, artefacts, and resources. Thus, despite their peripheral location, the national parks are prominently integrated into transnational interweaving spaces. These interdependencies affect the political and scientific dimensions of global nature conservation as well as the expansion of international tourism. This spatial configuration is repeatedly expressed by the most diverse actors when the difficult accessibility of the parks and thus the fragile connection to the center – especially to Buenos Aires – is addressed. On the other hand, the parks are characterized by a spatial difference between inside and outside, which is ultimately determined by the park boundaries. The interior of the park functions as a privileged field of intervention in which government practices are tested in the formation period of the national parks analyzed here, as are spatial materialization in the form of cities, transport infrastructure, and tree nurseries, while simultaneously enacting nature conservation rules. At the same time, a striking difference to the outside of the park remains. But the boundaries are not fixed. Rather, they can be understood as membranes that regulate permeability. For example, the park has a major influence on the development of tourism in the surrounding region.

Fourthly, the National Park regulates the relationship between nature and society on the basis of spatial difference. A whole network of specific governmental techniques, legal regulations and architectural institutions spans around special natural landscapes and unique natural monuments. The creation of national parks is a highly paradoxical operation. It is based on the separation of nature and culture, wilderness and civilization that Western thinking has created. At the same time, nature inscribes itself in the state via the national parks. The term park alone makes the paradoxical logic of the operation clear. "Wild nature" is to be civilized as a cultivated park under state control. The logic of the colonization of nature is thus inscribed in the bridgehead National Park.

In the first half of the 20th century, the national parks can be understood as bridgeheads of a new biopolitics, in which new government techniques, including their spatialization, emerge with regard to nature. Nature is regulated, controlled, parcelled out and disciplined, but also new knowledge and new images are produced, even a new imaginary of the nation. With these techniques, the national parks brought to an end the project of colonizing peripheral regions, which was accelerated in the last third of the 19th century. National parks advanced the urbanization, infrastructural development and economic use of peripheral regions. This included above all the designation of these areas as hotspots of a new tourism regime, while the model of agricultural colonization, which had been protected since the middle of the 19th century, came to an end. The national parks have an active population regulation system that is directed against indigenous groups – in northern Patagonia primarily against the Chilean Mapuche – and from the 1960s has increasingly followed the model of the "parks without people". In contrast, however, the presence of national and international tourists in the parks is increasing. Human voids are only small specifically designated protection zones. Numerous instruments and laws to regulate nature are enacted in the National Park bridgehead. In addition to regulations for the protection of natural resources, this also includes the active promotion of biological colonization. Particularly relevant is the distribution of salmonids and trees that extend far beyond the borders of the parks into the region.

If the close connection between national parks and colonization was thus worked out here, then this is not meant to mean that this connection would be an essential part of the parks. Rather, they can be understood as – to paraphrase Foucault – a decidedly heterogeneous ensemble of different, sometimes conflicting practices and techniques. In this sense, the genealogy of Argentine national parks is characterized by a transnational bricolage. Especially around 1913 and into the 1920s, there were quite different proposals for the conceptualization of the national parks – one has to think in particular of the proposals of Lucién Hauman and Carl Curt Hosseus – that had proposed alternative government techniques. Despite this heterogeneity, they did not move outside the idea of the national park as a bridgehead. However, nature's biopolitics were conceived differently.

Today's national parks in Argentina are a manifesto of these national park bridgeheads. The spatial-territorial structure of the parks has largely been preserved, also with regard to the relationship to the centers. The geopolitical imaginary of colonization is no longer relevant in today's discourse on national parks. Yet the traces of the explicit colonization strategy still shine through today. Neobiota, which were introduced through targeted biological colonization programs, have established themselves in the park and are the subject of fierce debates regarding the loss of biodiversity. The expropriated and robbed indigenous population highlights land rights and territorial claims in the parks. The park authorities, on the other hand, have been very hesitant in their response to the payment of the "historical debt" and have so far introduced only very limited models of the co-management.

The aesthetic-symbolic colonization of landscape also remains visible in the form of monuments and architecture. Graffiti in Puerto Iguazú or Mapuche symbols painted on the pavements of the central square in the Centro Cívico in Bariloche may correct these only slightly, while the *Rewe*, a pillar-like sacred altar of the Mapuche people, set up on the beach promenade in Bariloche are more likely to have an exoticizing and thus neocolonizing character. In terms of nature conservation, with the dimensions of landscape, natural monument, flora and fauna, the National Park, on the other hand, can boast many successes. The nature conservation measures took effect inside the parks, while the surrounding countryside suffered further ecological degradation.

In view of the contemporary multiple ecological crisis in the Anthropocene era, the question now arises as to whether the national parks could also serve as bridgeheads for a new ecological reconquest. Could regulatory techniques, legal regulations, the production of environmental knowledge, concepts, and architectural materializations be conceived of as national parks that could provide approaches to overcoming the multiple ecological crisis?

Various biologists have recently argued in this direction under the motto "the frontiers of wilderness" and called for the absolute protection of the last remaining wilderness areas. Exequiel Bustillo had still branded the positions oriented on the U.S.-American model of Wilderness as orthodox and clearly rejected them. Also in the current environmental historical debate, the concept of Wilderness has been sharply criticized and rejected due to its colonial implications and the conceptual separation of nature and culture. It is therefore all the more astonishing that it is experiencing a new, unexpected boom in the current anthropocene. According to the advocates of the new Wilderness movement, however, legal regulations – such as those in national parks and other protected areas – have proved to be highly effective for nature conservation. In this context, it is particularly important to move away from the insular structure of the protected areas and to establish broad ecological corridors between the parks – also across borders – that guarantee the circulation of biota. Contrary to Bailey Willis' utilitarian anthropocentric justification of national parks for the benefit of the national population, these nature conservation debates emphasize the planetary importance of parks for the vital stability of ecosystems, also with regard to climate change. Consequently, part of the scientific community in the current anthropocene is against any use, including tourism, of these last planetary wilderness areas (Watson et al. 2018).

While the idea of deserted parks is supported here, other ecosystem biological work points to the importance of indigenous peoples in preserving biodiversity. According to these findings, the proportion of vertebrates in indigenous areas in Canada, Australia and Brazil is as high, or even higher, than in protected areas (Schuster et al. 2019). This finding is a clear indication that the problem of ecological degradation is not caused by humans themselves, but by the way they are used. This should also be borne in mind in the anthropocene debate. Here, too, it is not man himself who overuses and litters the earth beyond planetary borders. Rather, the anthropocene is causally linked to the colonial expansion and formation of a capitalist world system that is geared to the progressive accumulation of wealth – which is also unequally distributed in historical uniqueness. With the tendency to commodify all areas of life, nature is also understood instrumentally as a resource to be exploited.

Particularly in Latin America, after the bursting of the financial speculation bubble, an ever-increasing, ruthless exploitation of natural resources and land-grabbing can be observed that follow colonial patterns (Svampa 2019). In view of this situation, any measure that deprives landscapes and natural areas of this logic of capitalist accumulation and land seizure is to be welcomed. National parks are certainly not a universal recipe for solving the multiple ecological crisis, which extends as far as ecocide. But they are certainly a true instrument against colonial land seizure.

From a historical perspective, the national parks also have a duty to work through their colonial debts as well as balance them. At the same time, however, rules, practices, institutions, knowledge production, norms and ethics have been established in the national parks that are based on models such as care and reparation and set limits to the excessive exploitation and colonization of nature.

Bibliography

Archives

Archivo General de la Nación, Buenos Aires (AGN)

- Legajo Exequiel Bustillo

Biblioteca Nacional, Buenos Aires (BN)

Biblioteca – Jardín Botánico Carlos Thays, Buenos Aires

Biblioteca Central y Centro de Documentación Perito Francisco Pascasio Moreno, Administración de Parques Nacionales, Buenos Aires (BM-APN)

Archivo, Administración de Parques Nacionales, Buenos Aires (APN)

Archivo Histórico de Arquitectura. Sociedad Central de Arquitectos, Buenos Aires

Biblioteca Perito Moreno, Museo de la Patagonia Francisco Pascasio Moreno, Bariloche (MP)

- Colección Emilio Frey (Colección Frey)
- Colección Francisco P. Moreno (Colección Moreno)

Ibero-Amerikanisches Archiv, Berlin (IAK)

State Archives of North Carolina

- Frank W. Bicknell Photograph Collection, 1905-1920, https://www.flickr.com/photos/north-carolina-state-archives/albums/72157606611483034 (Bicknell, PC)

Biodiversity Heritage Library, https://www.biodiversitylibrary.org/

Primary Sources

Ambrosetti, Juan B. 1903. *Colonias militares en Misiones*. Buenos Aires: Imprenta Martín Biedma.

Artayeta, Enrique Amadeo. 1945. *Biografía del perito Dr. Francisco P. Moreno*. Buenos Aires: Administración General de Parques Nacionales y Turismo.

Autran, Eugène. 1907. "Les Parcs Nationaux Argentins. Le Parc National de l'Iguazu. Le Parc National du Nahuel Huapi et sa florule." *Boletín del Ministerio de Agricultura* VII (1/2): 1-13.

Bayer, Osvaldo. 2020. "Desmonumentar". *Página/12*, 15.5.2010.

Berg, Carlos. 1877. "Enumeración de las plantas européas que se hallan como silvestres en la provincia de Buenos Aires y en patagonia". *Sociedad científica argentina* III: 182-206.

Berjman, Sonia. Ed. 2002. *Carlos Thays: sus escritos sobre jardines y paisajes*. Buenos Aires: Editorial Ciudad Argentina.

Bernárdez, Manuel. 1901. *De Buenos Aires al Iguazú*. Buenos Aires: La Nación.

Burmeister, Carlos. 1899. *Memoria sobre el territorio de Misiones*. Buenos Aires: Ministerio de Agricultura.

Bustillo, Exequiel. 1937. "Discurco pronunciado por el Dr. D. Exequiel Bustillo, el 20 de enero de 1937, con motive de la inauguración de la red de caminos en el Parque Nacional Nahuel Huapi". 20.1.1937.

———. 1965. "El problema de nuestra frontera patagónica. Conferencia en el ateneo de estudios de la Federación de Partidos de Centro". 21.12.1965.

———. 1972. *Huellas de un largo quehacer. Discursos, conferencias, artículos y publicaciones diversas*. Buenos Aires: Ediciones Depalma.

———. 1997 [1968]. *El despertar de Bariloche*. Buenos Aires: Editorial Sudamericana.

Cámara de diputados de la nación. 1934. *Diario de sessiones. Sessiones ordinarias*. Buenos Aires: 1321-1337.

Cámara de senadores de la nación. 1934. *Diario de sessiones*. Buenos Aires: 298-309; 631-643.

Cayuqueo, Pedro. 2018. "¿Los mapuche invasores chilenos?". *La Tercera*, 11.6.2018.

Conwentz, Hugo. 1904. *Die Gefährdung der Naturdenkmäler und Vorschläge zu ihrer Erhaltung. Denkschrift, dem Herrn Minister der geistlichen, Unterrichts- und Medizinal-Angelegenheiten überreicht*. Berlin.

———. 1915. "Naturschutzgebiete in Deutschland, Österreich und einigen anderen Ländern". *Zeitschrift der Gesellschaft für Erdkunde zu Berlin*: 29-51.

Devoto, Franco and Máximo Rothkugel. 1936. *Informe sobre los bosques del Parque nacional del Iguazú*. Buenos Aires: Ministerio de Agricultura.

Dirección de Parques Nacionales. 1936-1937. *Parque Nacional de Nahuel-Huapi*. Buenos Aires: Dirección de Parques Nacionales.
———. 1937. *Parque Nacional del Iguazú*. Buenos Aires: Dirección de Parques Nacionales.
———. 1937a. *Nuevos Parques Nacionales*. Buenos Aires: Dirección de Parques Nacionales.
———. 1938. *Obra Pública, cultural y turística realizada en los Parques Nacionales*. Buenos Aires.
———. 1938a. *Parque Nacional de Nahuel-Huapi. Historia, tradiciones y etnología*. Buenos Aires.
———. 1936. *Memoria correspondiente al año 1935*. Buenos Aires.
———. 1937. *Memoria correspondiente al año 1936*. Buenos Aires.
———. 1938. *Memoria correspondiente al año 1937*. Buenos Aires.
———. 1939. *Memoria correspondiente al año 1938*. Buenos Aires.
———. 1940. *Memoria correspondiente al año 1939*. Buenos Aires.
———. 1941. *Memoria correspondiente al año 1940*. Buenos Aires.
———. 1942. *Memoria correspondiente al año 1941*. Buenos Aires.
———. 1943. *Memoria correspondiente al año 1942*. Buenos Aires.
Doello Jurado, Martín. 1923. *Los proyectos de Parques Naturales en la región del Plata: Iniciativas para su realización*. Editor Soc. Argentina de Ciencias Naturales.
———. 1913. "Conveniencia de establecer un parque natural en los alrededores de Buenos Aires". *Physis* 1 (4): 200-206.
———. 1943. "Algo más sobre protección a la naturaleza". *Physis* 19 (53): 402-407.
FF.CC. 1928. *Parque Nacional del Sud. Lagos selvas, montañas*. Buenos Aires.
———. 1929. *Lagos andinos: Parque Nacional del Sud: Montañas, bosques, cascadas*. Buenos Aires.
Gambier, Marina. 2007. "Alejandro Bustillo". *La Nación*, 14.11.2007, https://www.lanacion.com.ar/arquitectura/alejandro-bustillo-nid961887.
Gandia, Enrique de. 1945. "La ciudad de los Césares". *Anales del Museo de la Patagonia* I: 101-120.
Gil, Martin. 1911. *Los árboles*. Buenos Aires: Consejo Nacional de Educación.

González Regalado, Tomás. 1945. "Peces de los Parques Nacionales Nahuel Huapi, Lanín y los Alerces". *Anales del Museo de la Patagonia* I: 121-138.

Groussac, Paul. 1920. *El viaje intellectual. Impresiones de naturaleza y arte*. Buenos Aires: JesúsMénéndez.

Hauman, Lucién. 1916. *La Forêt Valdivienne et ses Limites. Notes de Geographique Botanique*. Trabajos del Instituto de Botanica y Farmacologia. Buenos Aires.

———. 1922. "Para la protección de la naturaleza en la República Argentina". *Physis* 5 (22): 283-301.

———. 1928. *Les modifications de la flore argentine sous l'action de la civilisation. Essai de géobotanique humaine*. Bruxelles: M. Hayez, Imprimeur de l'Académie Royale de Belgique.

Hauman, Lucién and Alberto Castellanos. 1927. *Bibliografía botánica argentina. Especialmente para los años 1922-1926*. Buenos Aires: Imprenta y Casa Editora Coni.

Herman, Gonzalo. 2018. "Parques Nacionales: el motor del turismo argentino". *La Clarín*, 5.11.2018.

Hildebrandt, Hans and Otto Meiling. 1930/1931. *Guía del Nahuel Huapi y Parque Nacional del Sud.* Bariloche: Oficina de Turismo de San Carlos de Bariloche.

Hosseus, Carl Curt. 1915. *La vegetación del Lago Nahuel Huapí y sus montañas*. Instituto de Botánica y Farmacología: Buenos Aires.

———. 1916. *El proyectado Parque Nacional del Sud*. Buenos Aires: Dirección General de Agricultura y Defensa Agrícola.

———. 1926. "Recomendar la sanción de una ley o por expedición un decreto de poder ejecutivo de la nación, para crear un parque nacional central en la sierra de Córdoba". *Boletín de la Universidad Nacional de La Plata* X (7): 557-559.

———. 1928. "Ein neuer Naturschutzpark in Argentinien". *Naturschutz und Naturparke* 5: 92-43.

———. 2004. *Notas autobiográficas* (*Miscelánea* 100, Academia Nacional de Ciencias). Córdoba, Argentina: Academia Nacional de Ciencias.

International Office for the Protection of Nature. 1931. *The International Office for the Protection of Nature. Its origin, its programme, its organisation*. Bruxelles: IOPN.

IUCN – International Union for Conservation of Nature. 2008. *Guidelines for Applying Protected Area Management Categories*. Gland, Switzerland: IUCN.

Kirchhoff, Herbert. 1945. *Los Parques Nacionales Argentinios*. Buenos Aires: Editorial Guillermo Kraft.

Koutché, Vsevolod. 1942. *Estación Forestal de Puerto Anchorena, Isla Victoria. Su Organización y trabajos* (= *Boletín Forestal. Correspondiente al año 1941*). Buenos Aires: Ministerio de Agricultura.

La Prensa. 1935. "Fueron dictados dos decretos relacionados con la administración y vigilancia de los bosques fiscales". 21.11.1935.

Lebedeff, Nicolás. 1932. *Apuntes sobre reforestación de los terrenos desarbolados de la zona del lago Nahuel Huapi*. Ministerio de Agricultura, Dirección general de tierras: Buenos Aires.

———. 1942. "Informe preliminar sobre los estudios de los bosques de la reserva 'Lanín'". *Estudios del Ingeniero forestal Nicolas Lebedeff* (= *Boletin Forestal. Correspondiente a los años 1938-1939-1940*). Buenos Aires: Ministerio de Agricultura: 7-52.

———. 1942a. "Estudio de algunos bosques de la reserva 'Lanín'". *Estudios del Ingeniero forestal Nicolas Lebedeff* (= *Boletin Forestal. Correspondiente a los años 1938-1939-1940*). Buenos Aires: Ministerio de Agricultura: 53-124.

———. 1942b. "Apuntes sobre la reforestación artificial en la isla Victoria". *Estudios del Ingeniero forestal Nicolas Lebedeff* (= *Boletin Forestal. Correspondiente a los años 1938-1939-1940*). Buenos Aires: Ministerio de Agricultura: 205-219.

Ley No. 6721, 29.9.1909.

Ley, without number, 8.4.1922.

Ley de Parques Nacionales, No. 12103, 9.10.1934.

López Mañán, Julio. 1971 [1916]. *Tucumán antigüo. Anotaciones y documentos*. Buenos Aires: Universidad Nacional de Tucumán.

Marini, Tomás L. 1936. *Los salmónidos en nuestro Parque Nacional de Nahuel Huapi. Comentarios a los trabajos de Piscicultura realizados en nuestro país*. Buenos Aires: Ministerio de Agricultura – División Piscicultura.

Massart, Jean. 1912. *Pour la conservation de la nature en Belgique*. Bruxelles: Lamertin.

Ministro del Interior. 1896. *Memoria del Ministro del Interior ante el Congreso Nacional, 1895*. Tomo III. Buenos Aires.

Molfino, José. 1928. "Nuevas adiciones a la flora fanerogámica adventicia de la Argentina". *Revista Fac. Agron. Univ. La Plata* 18 (2): 139-156.

Morales, Emilio B. 1929. *Iguazú. Cataratas y ruinas*. Buenos Aires: Talleres gráficos Argentinos.

Moreno, Francisco P. 1917. "Tierras y colonias". *La Unión*, 12.5.1917.

OAS. 1940. *Convention on Nature Protection adn Wild Life Preservation in the Western Hemisphere*. Washington D.C. http://www.oas.org/juridico/english/treaties/c-8.html.

Physis. Revista de la Sociedad Argentina de Ciencias Naturales. Vol. III. 1916-1917.

———. Vol. IV. 1918-1919.

———. Vol. V. 1921-1922.

Polgar, Estebán. 1964. "El ingeniero Emilio E. Frey, colaborador y amigo del perito Francisco P. Moreno". *Diario Rio Negro*. 27.6.1964.

Pusso, Santiago. 1912. *Viajes por mi tierra, Al Iguazú – a Nahuel Huapi – por las costas del sur*. Barcelona: Casa Editorial Maucci.

Radovich, Juan Carlos and Alejandro Balazote. 2009. "El pueblo mapuche contra la discriminación y el etnocidio". *Historia de las familias mapuche Lof Paichil Antriao y Lof Quintriqueo de la margen Norte del lago Nahuel Huapi*. Ed. Gerardo Ghiolgi. Villa la Angostura, Argentina: Archivos del Sur: 35-51.

Ramos Mexía, Ezequiel. 1921. *La colonización oficial y la distribución de las tierras públicas*. Buenos Aires: Ferrari.

———. 1936. *Mis Memorias 1853-1935*. Buenos Aires, Librería y Editorial La Facultad.

Reclus, Elisée. 1894. *Nouvelle Géographie Universelle. Vol. 19. Amérique du Sud*. Paris: Hachette.

Reed, Captain. 1949. *Fishing in Nahuel Huapi*. Buenos Aires: Administración General de Parques Nacionales y Turismo.

Roosevelt, Theodore. 1914a. "Mi viaje a Sud América". *La Prensa*, 1.6.1914.

———. 1914b. "Mi viaje a Sud América". *La Prensa*, 2.6.1914.

———. 1916. *A Book-Lover's Holidays in the Open*. New York: Charles Scribner's Sons.

Ruíz Moreno, Isidoro. 1933. *En el Nahuel Huapi, en el año 1913 con el ex presidente de los Estados Unidos de Norte América, Don Teododo Roosevelt. Breve Crónica de la visita de Teododo Roosevelt a la región del Nahuel Huapi*. Buenos Aires.

Sarasin, Paul. 1914. *Ueber die Aufgaben des Weltnaturschutzes. Denkschrift gelesen an der Delegiertenversammlung zur Weltnaturschutzkommission in Bern am 18. November 1913.* Basel: Verlag von Helbig & Lichtenbahn.

Tercer Congreso Científico Pan-Americano. 1924. *Constitución y Programas*. Lima.

———. 1938. *Acta final*. Lima.

Thays, Carlos (son). 1918. *Consideraciones sobre la Formación del Parque Nacional del Sud / Nahuel Huapi*. Tesis Facultad de Agronomía y Veterenaria, Universidad de Buenos Aires.

Thays, Carlos. 1902. "La excursion del M. Thays al Iguazú. Króncia de su viaje". *Caras y Caretas*. 31.5.1902: 42-45.

———. 1912. *El Parque Nacional Iguazú*. Buenos Aires (unpublished). Reprinted in Berjman (2002: 323-337).

———. 1913. *Les Fôrets naturelles de la République Argentine*. Touring Club de France. París. Reprinted in Berjman (2002: 339-363).

Tobal, Miguel Angel. 1912. *Anales de la Sociedad Forestal Argentina*. [S.I.]: Sociedad Forestal Argentina.

———. 1923. *El problema del árbol: memoria de la obra realizada por la Sociedad Forestal Argentina.* Buenos Aires: Talleres Grafícos Argentinos L.J. Rosso.

———. 1950. *La difusión del arbol en nuestro país y uno de sus grandes precursores*. Buenos Aires: El Ateneo.

Tulian, E. A. 1910. "Acclimatization of American Fishes in Argentina". *Bulletin of the Bureau of Fisheries* 28. Washington, D.C.: U.S. Government Printing Office: 957-965.

Willis, Bailey. 1913. *El Parque Nacional del Sud. Dirección General de Agricultura y Defensa Agrícula*. Buenos Aires.

———. 1943. *El norte de la Patagonia. Historia de la comisión de estudios hidrológicos del Ministerio de Obras Públicas – 1911-1914.* Buenos Aires: Dirección de Parques Nacionales.

———. 1947. *A Yanqui in Patagonia*. Stanford: Stanford University Press.

———. 1988. *El norte de la Patagonia*. Buenos Aires: Comisión de estudios hidrológicos.

Secondary Literature

Allan, J. R. et al. 2017. "Temporally inter-comparable maps of terrestrial wilderness and the Last of the Wild". *Scientific Data* 4: 170-187.

Bandieri, Susana. 2009. "Pensar una Patagonia con dos océanos. El proyecto de desarrollo de Ezequiel Ramos Mejía". *Quinto Sol – Revista de Historia Regional* 13: 47-72.

Barcina, Florencia. Ed. 2007. *Ernesto Estrada: el arquitecto frente al paisaje*. Buenos Aires: CEDODAL.

Berjman, Sonia. 2009. *Carlos Thays. Un jardinero francés en Buenos Aires*. Embajada de Francia: Buenos Aires.

———. 2002a. "Introducción". *Carlos Thays: sus escritos sobre jardines y paisajes.* Ed. Sonia Berjman. Buenos Aires: Editorial Ciudad Argentina: 15-39.

Berjman, Sonia and Ramón Gutiérrez. 1988. *Patrimonio cultural y patrimonio natural: La arquitectura en los Parques Nacionales Nahuel Huapi e Iguazú.* Buenos Aires: Editorial del Instituto Argentino de Investigaciones de Historia de la Arquitectura y del Urbanismo.

Bessera, Eduardo Miguel. 2008. *Políticas de Estado en la Norpatagonia Andina. Parques Nacionales, desarrollo turístico y consolidación de la frontera. El caso de San Carlos de Bariloche (1934-1955).* Tesis de Licenciatura. Universidad Nacional de Comahue.

Biedma, Juan M. 2003. *Crónica histórica del lago Nahuel Huapí*. Bariloche: Ediciones Caleuche.

Blanc, Jacob and Federico Freitas. Ed. 2018. *Big Water. The Making of the Borderlands between Brazil, Argentina, and Paraguay*. Tucson: The University of Arizona Press.

Brinkley, Douglas. 2010. *The Wilderness Warrior. Theodore Roosevelt and the Crusade for America*. New York: Harper Perennial.

Briones, Claudia. 2000. "Contemporary Criticisms of the Received Wilderness Idea". *Wilderness Science in a Time of Change Conference. Volume 1: Changing Perspectives and Future Directions. Missoula, Montana, May 23-27, 1999* (Rocky Mountain Research Station,

Proceedings RMRS-P-15-VOL-1). Fort Collins, CO: Rocky Mountain Research Station: 24-31 (https://www.fs.usda.gov/treesearch/pubs/6415).

Carreras Doallo, Ximena A. 2015. "El Parque Nacional Iguazú en clave soberana: el rol de Thays en la imagen de la Argentina". *Anuario del Centro de Estudios Históricos "Prof. Carlos S. A. Segreti" Córdoba (Argentina)* 15 (15): 167-184.

Clifford, James. 1997. *Routes. Travel and Translation in the Late Twentieth Century*. London: Harvard UP.

Crespo, Carolina. 2021. "Memories in Displacement in the Public Space. The Monuments of Juana Azurduy and Christopher Columbus in Argentina". *Cherishing the Past, Envisioning the Future. Entangled Practises of Heritage and Utopia in the Americas*. Ed. Olaf Kaltmeier, Mirko Petersen, Wilfried Raussert and Julia Roth. Trier, New Orleans: WVT/UNO: 101-122.

Crosby, Alfred W. 2004. *Ecological Imperialism. The Biological Expansion of Europe*. 2. ed., New York: Cambridge UP.

———. 1972. *The Columbian Exchange: Biological and Social Consequences of 1492.* Santa Barbara: Greenwood.

Davis, Mark A. 2006. "Invasion biology 1958-2005: The pursuit of science and conservation". *Conceptual ecology and invasions biology: reciprocal approaches to nature.* Ed. Marc William Cadotte, Sean M. McMahon and Tadashi Fukami. Dordrecht: Springer: 35-64.

Encyclopedia Britannica. 2019. "National Park." https://www.britannica.com/science/national-park.

Feiber, Albert. 2016. "Filiale von Berlin. Der Obersalzberg im Dritten Reich". *Die tödliche Utopie. Bilder, Texte, Dokumente, Daten zum Dritten Reich.* Ed. Volker Dahm, Albert A. Feiber, Hartmut Mehringer and Horst Möller. München/Berlin: Verlag Dokumentation Obersalzberg im Institut für Zeitgeschichte: 53-187.

Ferrero, Brián and Luz Pyke. 2017. "Un Parque en la frontera: El rol del Parque Nacional Iguazú en la constitución de la frontera internacional entre Argentina y Brasil". *Faces da fronteira: Desafios e Perspectivas de Regiões Lindeiras.* Vol. II. Ed. I. Cavlak, J. L. Rebelo Porto and A. E Noronha. UNIFAP: Macapá: 99-126.

Ferrero, Brian and Luz Pyke. 2016. "Naturaleza y Frontera. El Parque Nacional Iguazú y el proceso de consolidación del Estado argentino en la frontera argentino-brasileña (1880-1934)". *Sociedad y Discurso* 28: 135-167.

Foucault, Michel. 1978. *Dispositive der Macht. Über Sexualität, Wissen und Wahrheit*. Berlin: Merve.

Freitas, Federico. 2014. "A Park for the Borderlands: The Creation of the Iguaçu National Park in Southern Brazil, 1880-1940". *Revista de Historia Iberoamericana* 7 (2): 65-88.

———. 2018. "Argentinizing the Border. Conservation and Colonization in the Iguazú National Park". *Big Water. The Making of the Borderlands between Brazil, Argentina, and Paraguay*. Ed. Jacob Blanc and Federico Freitas. Tucson: The University of Arizona Press: 105-130.

Frondizi, Arturo. 1964. *Breve historia de un yanqui que proyectó industrializar la Patagonia*. Buenos Aires: Ediciones Cen.

Gissibl, Bernhard, Sabine Höhler and Patrick Kupper. Eds. 2012. *Civilizing Nature. National Parks in Global Historical Perspective*. New York: Berghahn Books.

Gissibl, Bernhard, Sabine Höhler and Patrick Kupper. 2012a. "Introduction: Towards a Global History of National Parks". *Civilizing Nature. National Parks in Global Historical Perspective*. New York: Berghahn Books: 1-27.

Gutiérrez, Ramon. 2002. "Buenos Aires, a Great European City". *Planning Latin America's Capital Cities*. Ed. Arturo Almandoz. London: Routledge: 45-74.

———. 2017. *Ernesto de Estrada. El arquitecto frente al paisaje*. Buenos Aires: Centro de Documentación de Arquitectura Latinoamericana.

Hardenberg, Wilko Graf von, Matthew Kelly, Claudia Leal and Emily Wakild. 2017. "Introduction". *The Nature State. Rethinking the History of Conservation*. Ed. Wilko Graf von Hardenberg, Matthew Kelly, Claudia Leal and Emily Wakild. New York: Routledge: 1-16.

Hirsch, Heidi et al. 2011. "Tracking the origin of invasive *Rosa rubiginosa* populations in Argentina". *International Journal of Plant Sciences* 172 (4): 530-540.

Illies, Florian. 2012. *1913. Der Sommer des Jahrhunderts*. Frankfurt a.M.: Fischer.

Kaltmeier, Olaf. 2004. *¡Marichiweu! – Zehnmal werden wir siegen! Eine Rekonstruktion der aktuellen Mapuche-Bewegung in Chile aus der Dialektik von Herrschaft und Widerstand seit der Conquista.* Münster: Edition ITP-Kompass.

———. 2011. "Selling EthniCity: Urban Cultural Politics in the Americas from the Conquest to Contemporary Consumer Societies". *Selling EthniCity: Urban Cultural Politics in the Americas.* Ed. Olaf Kaltmeier. Farnham: Ashgate: 1-20.

———. 2019. "Invidious Comparison and the New Global Leisure Class: On the Refeudalization of Consumption in the Old and New Gilded Age". *Fiar – Forum for Inter-American Reseach* 12 (1): 29-42.

Karpinski, Cezar. 2013. "A construção de uma "Maravilha": Brasil e Argentina e as disputas pela paisagem das Cataratas do Iguaçu (1880-1914)". http://www.snh2013.anpuh.org/resources/anais/27/1364428427_ARQUIVO_Artigocompleto.pdf.

Keiter, Robert B. 2013. *To Conserve Unimpaired: The Evolution of the National Park Idea.* Washington D.C.: Island Press.

Kinsey, Darin. 2006. "Seeding the Water as the Earth: The Epicenter and Peripheries of a Western Aquacultural Revolution". *Environmental History* 11 (3): 527-566.

———. 2008. *Fashioning a Freshwater Eden. Elite anglers, fish culture, and state development of Québec's sport fishery.* Dissertation, Université du Québec à Trois-Rivières.

Kupper, Patrick. 2012. "Translating Yellowstone: Early European National Parks, Weltnaturschutz and the Swiss Model." *Civilizing Nature: National Parks in Global Historical Perspective.* Ed. Bernhard Gissibl, Sabine Höhler and Patrick Kupper. New York: Berghahn Books: 123-139.

Levisman, Martha. 2010. *Bustillo en Patagonia.* Buenos Aires: Arca.

Lolich, Liliana. 2007. "Arquitectura de los Parques Nacionales (1934-1955)". *Ernesto de Estrada. El arquitecto frente al paisaje.* Ed. Ramón Gutiérrez. Buenos Aires: Centro de Documentación de Arquitectura Latinoamericana: 79-94.

Losada, Leandro. 2007. "La alta sociedad y la política en la Buenos Aires del novecientos: la sociabilidad distinguida durante el orden conservador (1880-1916)". *Entrepasados* 31: 81-96.

Macchi, Patricio and Pablo Vigliano. 2014. "Salmonid introduction in Patagonia: the ghost of past, present and future management". *Ecología Austral* 24: 162-172.

Macchi, Patricio, Miguel Pascual and Pablo Vigliano. 2007. "Differential piscivory of the native Percichthys trucha and exotic salmonids upon the native forage fish Galaxias maculatus in Patagonian Andean lakes". *Limnologica* 37: 76-87.

Mann, Charles C. 2011. *1493*: *Uncovering the New World Columbus created.* New York: Vintage.

———. 2019. "Columbian Exchange". *The Routledge Handbook to the History and Society of the Americas*. Ed. Olaf Kaltmeier, Josef Raab, Mike Foley, Mario Rufer, Stefan Rinke and Alice Nash. New York: Routledge: 67-74.

Matossian, Brenda. 2012. "Modelos de desarrollo, poblamiento y frontera. El caso del Parque Nacional Nahuel Huapi". *Estudios sociales Contemporáneos* 5 (6): 67-84.

McCormick, John 1991. *Reclaiming Paradise: The Global Environmental Movement.* Bloomington: Indiana UP.

Meyer, John W. et al. 1997. "The Structuring of a World Environmental Regime, 1870-1990". *International Organization* 51 (4): 623-651.

Miller Klubock, Thomas. 2014. *La Frontera: Forests and Ecological Conflict in Chile's Frontier Territory*. Durham: Duke University Press.

Morgenfeld, Leandro Ariel. 2007. "Argentina frente a Estados Unidos en las conferencias panamericanas de los años 30". *Ciclos en la historia, la economía y la sociedad* 16 (31/32): 193-217.

Mosena, Alexander, Wolfram Beyschlag and Tom Steinlein. 2018. "Reconstructing the historical spread of non-native plants in the North American West from herbarium specimens". *Flora* 242: 45-52.

Municipio de San Carlos de Bariloche. n.y. *Isla Victoria. Datos históricos*. http://www.bariloche.gov.ar/upload/SECRETARIA%20DE%20TURISMO/371_Isla%20Victoria.pdf.

Navarro Floria, Pedro. 2011. "Las viejas fronteras revisitadas: problematizando la formación territorial de los bordes de los Estados-nación latinoamericanos a través del caso de la Norpatagonia Argentina". *Antíteses* 4 (8): 427-454.

———. 2010. "Procesos de resignificación, desarrollos locales y formación territorial en los territories nacionales. El caso de la "Suiza argentina" entre los siglos XIX y XX". *Anuario IEHS* 25: 45-71.

Navarro Floria, Pedro and Laila Vejsberg. 2009. "El proyecto turístico barilochense antes de Bustillo. Entre la prehistoria del Parque Nacional Nahuel Huapi y el desarrollo local". *Estudios y Perspectivas en Turismo* 18 (4): 414-433.

Navarro Floria, Pedro and Laila Vejsberg. 2010. "El tourismo. Entre la actividad económica y el derecho social. El Parque Nacional Nahuel Huapi, Argentina, 1934-1955". *Estudios y Perspectivas en Turismo* 19 (6): 930-945.

Nuñez, Martín Andrés, Maria Andrea Relva and Daniel Simberloff. 2010. "Introduced deer affect native plant communities and facilitate non-native tree species in a temperate forests in Patagonia, Argentina". *Biological Invasions* 12: 303-311.

Ortemberg, Pablo. 2017. "Panamericanismo, hispanoamericanismo y nacionalismo en los festejos identitarios de América Latina, 1880-1920". *Anuario IEHS* 32 (1): 99-110.

Osborne, Michael. 2000. "Acclimatizing the World: A History of the Paradigmatic Colonial Science". *Osiris* 15: 135-151.

Osterhammel, Jürgen. 2009. *Die Verwandlung der Welt. Eine Geschichte des 19. Jahrhunderts*. München: C.H. Beck.

Piglia, Melissa. 2014. *Autos, rutas y turismo. El Automóvil Club Argentino y el estado*. Buenos Aires: Siglo XXI.

Policastro, Carolina and Ana Marks. 2008. "Pobladores mapuches en el Nahuel Huapi: de la exclusión a la adscripción". *IX Congreso Argentinode Antropología Social*. Posadas: Facultad de Humanidades y Ciencias Sociales – Universidad Nacional de Misiones. http://cdsa.aacademica.org/000-080/470.pdf.

Potapov, Peter, Matthew C. Hansen, Lars Laestadius, Svetlana Turubanova, Alexey Yaroshenko, Christoph Thies, Wynet Smith, Ilona Zhuravleva, Anna Komarova, Susan Minnemeyer and Elena Esipova. 2017. "The last frontiers of wilderness: Tracking loss of intact forest landscapes from 2000 to 2013". *Science Advances* 3 (1).

Prat, Adela. 2017. "La CSJN Declara Inconstitucionales Dos Leyes Provinciales. En Cuanto Disponen La Creación Del "Parque Provincial Del Río Iguazú" Sobre un Establecimiento De Utilidad Nacional: El

"Parque Nacional Iguazú" Sometido a Jurisdicción Del Estado Nacional". Blog Legal de la Doctora Adela M. Prat. https://www.adelaprat.com/2017/08/la-csjn-declara-inconstitucionales-dos-leyes-provinciales-en-cuanto-disponen-la-creacion-del-parque-provincial-del-rio-iguazu-sobre-un-establecimiento-de-utilidad-nacional-el-parque-nacional-i/.

Ramos, Jorge. 1993. "Alejandro Bustillo: de la Hélade a la Pampa". *Seminario de Crítica* 34. http://www.iaa.fadu.uba.ar/publicaciones/critica/0034.pdf.

Rinke, Stefan. 2015. *Im Sog der Katastrophe. Lateinamerika und der Erste Weltkrieg*. Frankfurt a.M.: Campus.

Ruffini, Martha E. 2007. "La consolidación inconclusa del Estado: los Territorios Nacionales, gobernaciones de provisionalidad permanente y ciudadanía política restringida (1884-1955)". *Revista SAAP* 3 (1): 81-101.

Runte, Alfred. 1979. *National Parks. The American Experience*. Lincoln: University of Nebraska Press.

Scarzanella, Eugenia. 2002. "Las bellezas naturales y la nación: Los parques nacionales en Argentina en la primera mitad del siglo XX". *Revista Europea de Estudios Latinoamericanos y del Caribe* 73: 5-21.

Schreuder, F.-G. 1969. "Zur Klassifizierung der Anthropochoren". *Vegetatio* 16 (5/6): 225-238.

Richard Schuster, Ryan R. Germain, Joseph R. Bennett, Nicholas J. Reo and Peter Arcese. 2019. "Vertebrate biodiversity on indigenous-managed lands in Australia, Brazil, and Canada equals that in protected areas". *Environmental Science & Policy* 101: 1-6.

Sellars, Richard West. 2009. *Preserving Nature in the National Parks. A History with a New Preface and Epilogue*. New Haven: Yale University Press.

Sennett, Richard. 1991. *The Conscience of the Eye. The Design and Social Life of Cities*. New York: Faber and Faber.

Sheinin, David. 1998. "Wilderness and Pan American Preservation, 1910-1948". *Latin American Studies Association, XXI International Congress, Chicago, Illinois, 24-26 September 1998*. http://biblioteca.clacso.edu.ar/ar/libros/lasa98/Sheinin.pdf.

Silvestri, Graciela. 2011. *El lugar común. Una historia de las figuras de paisaje del Río de la Plata*. Buenos Aires: Edhasa.

Simberloff, Daniel, Relva, Maria Andrea and Martín Nuñez. 2002. "Gringos en el Bosque: Introduced Tree Invasion in a Native Nothofagus/Austrocedrus Forest". *Biological Invasions* 4: 35-53.

Staropoli, Laura Liliana. 2010. *Los documentos históricos de los pobaldores y comunidades vinculados a los Parques Nacionales en el archive de la APN*. Buenos Aires: APN.

Steffen, Will, Wendy Broadgate, Lisa Deutsch, Owen Gaffney and Cornelia Ludwig. 2015. "The trajectory of the Anthropocene: The Great Acceleration". *The Anthropocene Review* 2 (1): 81-98.

Svampa, Maristella. 2019. *Las fronteras del neo-extractivismo*. Bielefeld, Guadalajara: BiUP, UdeG.

Symoens, Jean-Jacques. 1998. "Un botánico belga en Argentina: Lucien Hauman". *En los deltas de la memoria: Bélgica y Argentina en los siglos XIX y XX*. Ed. Bart de Groof, Patricia Geli, Eddy Stols and Guy Van Beeck. Leuven: Leuven University Press: 143-145.

Trentini, Florencia. 2015. *Pueblos indígenas y áreas protegidas: Procesos de construcción de identidades y territorialidades en el comanejo del Parque Nacional Nahuel Huapi*. Doctoral thesis. Universidad de Buenos Aires.

———. 2011. "Entre la conservación y la legitimidad: El caso de la comunidad Maliqueo y el Parque Nacional Nahuel Huapi". *Kula Antropólogos del Atlántico Sur* 4: 61-75.

Tyrell, Ian. 2015. *Crisis of the Wastefull Nation. Empire and Conservation in Theodore Roosevelt's America*. Chicago: The University of Chicago Press.

Vallejos, Jorge Emanuel and Alejandra Beatriz González. 2014. *Selva, Tierra y Agua. 80 años del Parque Nacional Iguazú (1934-2014)*. Buenos Aires: APN.

Valverde, Sebastián. 2011. "Demandas territoriales del pueblo mapuche en área Parques Nacionales". *Avá – Revista de Antropología* 17: 69-83.

Wakild, Emily. 2017. "Protecting Patagonia. Science, conservation and the pre-history of the nature-state on a South American frontier, 1903-1934." *The Nature State. Rethinking the History of Conserva-*

tion. Ed. Wilko Graf von Hardenberg, Matthew Kelly, Claudia Leal and Emily Wakild. New York: Routledge: 37-54.

Watson, James E.M., Oscar Venter, Jasmine Lee, Kendall R. Jones, John G. Robinson, Hugh P. Possingham and James R. Allan. 2018. “Protect the last of the wild”. *Nature* 563: 27-30.

Wilde, Guillermo. 2007. “De la depredación a la conservación. Génesis y evolución del discurso hegemónico sobre la selva misionera y sus habitantes”. *Ambiente & Sociedade Campinas* X (1): 87-106.

Williams, Michael. 2006. *Deforesting the Earth: From Prehistory to Global Crisis. An Abridgment*. Chicago: University of Chicago Press.

Zarrilli, Adrián. 2008. “Bosques y agricultura: Una mirada a los límites históricos de sustentabilidad de los bosques argentinos en un contexto de la explotación capitalista en el siglo XX”. *Revista Luna Azul* 26: 87-106.

Zimmermann, Christiane M. Ritz, Volker Wissemann, Henrik von Wehrden, Daniel Renison, Karsten Wesche, Erik Welk and Isabell Hensen. 2011. “Tracking the Origin of Invasive Rosa rubiginosa Populations in Argentina”. *International Journal of Plant Sciences* 172 (4): 530-540.

Zusman, Perla. 2009. “Panamericanismo y Nacionalismo en torno al viaje de Teodoro Roosevelt al norte de la Patagonia (1913)”. *XII Jornadas Interescuelas/Departamentos de Historia*. Departamento de Historia, Facultad de Humanidades y Centro Regional Universitario Bariloche. Universidad Nacional del Comahue, San Carlos de Bariloche. http://cdsa.aacademica.org/000-008/202.pdf.

INTER-AMERICAN STUDIES
Cultures – Societies – History

ESTUDIOS INTERAMERICANOS
Culturas – Sociedades – Historia

This interdisciplinary series examines national and transnational issues in the cultures, societies, and histories of the Americas. It creates a forum for a critical academic dialogue between North and South, promoting an inter-American paradigm that shifts the scholarly focus from methodological nationalism to the wider context of the Western Hemisphere.

Vol. 1
Raab, Josef, Sebastian Thies, and Daniela Noll-Opitz, eds. *Screening the Americas: Narration of Nation in Documentary Film / Proyectando las Américas: Narración de la nación en el cine documental*. 2011. 470 pp.

WVT Wissenschaftlicher Verlag Trier	ISBN 978-3-86821-331-7	€ 29,50
Bilingual Press / Editorial Bilingüe	ISBN 978-1-931010-83-2	$ 29.50

Vol. 2
Raussert, Wilfried, and Michelle Habell-Pallán, eds. *Cornbread and Cuchifritos: Ethnic Identity Politics, Transnationalization, and Transculturation in American Urban Popular Music*. 2011. 292 pp.

WVT Wissenschaftlicher Verlag Trier	ISBN 978-3-86821-265-5	€ 29,50
Bilingual Press / Editorial Bilingüe	ISBN 978- 1-931010-80-1	$ 29.50

Vol. 3
Butler, Martin, Jens Martin Gurr, and Olaf Kaltmeier, eds. *EthniCities: Metropolitan Cultures and Ethnic Identities in the Americas*. 2011. 268 pp.

WVT Wissenschaftlicher Verlag Trier	ISBN 978-3-86821-310-2	€ 29,50
Bilingual Press / Editorial Bilingüe	ISBN 978-1-931010-81-8	$ 29.50

Vol. 4
Gurr, Jens Martin, and Wilfried Raussert, eds. *Cityscapes in the Americas and Beyond: Representations of Urban Complexity in Literature and Film*. 2011. 300 pp.

WVT Wissenschaftlicher Verlag Trier	ISBN 978-3-86821-324-9	€ 29,50
Bilingual Press / Editorial Bilingüe	ISBN 978-1-931010-82-5	$ 29.50

Vol. 5

Kirschner, Luz Angélica, ed. *Expanding* Latinidad*: An Inter-American Perspective*. 2012. 292 pp.

WVT Wissenschaftlicher Verlag Trier	ISBN 978-3-86821-309-6	€ 29,50
Bilingual Press / Editorial Bilingüe	ISBN 978-1-931010-84-9	$ 29.50

Vol. 6

Raussert, Wilfried, and Graciela Martínez-Zalce, eds. *(Re)Discovering 'America': Road Movies and Other Travel Narratives in North America / (Re)Descubriendo 'América': Road movie y otras narrativas de viaje en América del Norte*. 2012. 252 pp.

WVT Wissenschaftlicher Verlag Trier	ISBN 978-3-86821-384-3	€ 29,50
Bilingual Press / Editorial Bilingüe	ISBN 978-1-931010-91-7	$ 29.50

Vol. 7

Kaltmeier, Olaf, ed. *Transnational Americas: Envisioning Inter-American Area Studies in Globalization Processes*. 2013. 278 pp.

WVT Wissenschaftlicher Verlag Trier	ISBN 978-3-86821-415-4	€ 29,50
Bilingual Press / Editorial Bilingüe	ISBN 978-1-931010-92-4	$ 29.50

Vol. 8

Raab, Josef, and Alexander Greiffenstern, eds. *Interculturalism in North America: Canada, the United States, Mexico, and Beyond*. 2013. 312 pp.

WVT Wissenschaftlicher Verlag Trier	ISBN 978-3-86821-460-4	€ 29,50
Bilingual Press / Editorial Bilingüe	ISBN 978-1-931010-99-3	$ 29.50

Vol. 9

Raab, Josef, ed. *New World Colors: Ethnicity, Belonging, and Difference in the Americas*. 2014. 418 pp.

WVT Wissenschaftlicher Verlag Trier	ISBN 978-3-86821-461-1	€ 29,50
Bilingual Press / Editorial Bilingüe	ISBN 978-1-939743-00-8	$ 39.50

Vol. 10

Roth, Julia. *Occidental Readings, Decolonial Practices: A Selection on Gender, Genre, and Coloniality in the Americas*. 2014. 284 pp.

WVT Wissenschaftlicher Verlag Trier	ISBN 978-3-86821-446-8	€ 26,50
Bilingual Press / Editorial Bilingüe	ISBN 978-1-939743-07-7	$ 32.50

Vol. 11

Thies, Sebastian, Gabriele Pisarz-Ramirez, and Luzelena Gutiérrez de Velasco, eds. *Of Fatherlands and Motherlands: Gender and Nation in the Americas / De Patrias y Matrias: Género y nación en las Américas*. 2015. 344 pp.

WVT Wissenschaftlicher Verlag Trier	ISBN 978-3-86821-528-1	€ 29,50
Bilingual Press / Editorial Bilingüe	ISBN 978-1-939743-08-4	$ 39.50

Vol. 12
Fuchs, Rebecca. *Caribbeanness as a Global Phenomenon: Junot Díaz, Edwidge Danticat, and Cristina García*. 2014. 298 pp.

WVT Wissenschaftlicher Verlag Trier	ISBN 978-3-86821-533-5	€ 26,50
Bilingual Press / Editorial Bilingüe	ISBN 978-1-939743-09-1	$ 32.50

Vol. 13
Andres, Julia. *¡Cuéntame algo! – Chicana Narrative Beyond the Borderlands*. 2015. 202 pp.

WVT Wissenschaftlicher Verlag Trier	ISBN 978-3-86821-569-4	€ 25,00
Bilingual Press / Editorial Bilingüe	ISBN 978-1-939743-11-4	$ 28.50

Vol. 14
Hertlein, Saskia. *Tales of Transformation: Emerging Adulthood, Migration, and Ethnicity in Contemporary American Literature*. 2014. 228 pp.

WVT Wissenschaftlicher Verlag Trier	ISBN 978-3-86821-570-0	€ 25,00
Bilingual Press / Editorial Bilingüe	ISBN 978-1-939743-10-7	$ 31.50

Vol. 15
Raab, Josef, and Saskia Hertlein, eds. *Spaces – Communities – Discourses: Charting Identity and Belonging in the Americas*. 2016. 382 pp.

WVT Wissenschaftlicher Verlag Trier	ISBN 978-3-86821-590-8	€ 29,50
Bilingual Press / Editorial Bilingüe	ISBN 978-1-939743-13-8	$ 39.50

Vol. 16
Mehring, Frank, ed. *The Mexico Diary: Winold Reiss between Vogue Mexico and Harlem Renaissance. An Illustrated Trilingual Edition with Commentary and Musical Interpretation* (includes color plates and audio CD). 2016. 244 pp.

WVT Wissenschaftlicher Verlag Trier	ISBN 978-3-86821-594-6	€ 29,50
Bilingual Press / Editorial Bilingüe	ISBN 978-1-939743-14-5	$ 39.50

Vol. 17
Raussert, Wilfried, Brian Rozema, Yolanda Campos, and Marius Littschwager, eds. *Key Tropes in Inter-American Studies: Perspectives from the* forum for inter-american research (fiar). 2015. 374 pp.

WVT Wissenschaftlicher Verlag Trier	ISBN 978-3-86821-627-1	€ 29,50
Bilingual Press / Editorial Bilingüe	ISBN 978-1-939743-16-9	$ 39.50

Vol. 19
Rehm, Lukas, Jochen Kemner, and Olaf Kaltmeier, eds. *Politics of Entanglement in the Americas: Connecting Transnational Flows and Local Perspectives*. 2017. 226 pp.

WVT Wissenschaftlicher Verlag Trier	ISBN 978-3-86821-675-2	€ 27,50
Bilingual Press / Editorial Bilingüe	ISBN 978-1-939743-17-6	$ 32.50

Vol. 20
Britt Arredondo, Christopher. *Imperial Idiocy: A Reflection on Forced Displacement in the Americas*. 2017. 194 pp.
WVT Wissenschaftlicher Verlag Trier ISBN 978-3-86821-706-3 € 26,50
Bilingual Press / Editorial Bilingüe ISBN 978-1-939743-20-6 $ 30.00

Vol. 21
Schemien, Alexia. *Of Virgins, Curanderas, and Wrestler Saints: Un/Doing Religion in Contemporary Mexican American Literature*. 2018. 218 pp.
WVT Wissenschaftlicher Verlag Trier ISBN 978-3-86821-724-7 € 27,50
Bilingual Press / Editorial Bilingüe ISBN 978-1-939743-22-0 $ 32.50

Vol. 22
Fulger, Maria Diana. *The Cuban Post-Socialist Exotic: Contemporary U.S. American Travel Narratives about Cuba.* 2020. 266 pp.
WVT Wissenschaftlicher Verlag Trier ISBN 978-3-86821-769-8 € 32,50
Bilingual Press / Editorial Bilingüe ISBN 978-1-939743-27-5 $ 36.00

Vol. 24
Kaltmeier, Olaf, Mirko Petersen, Wilfried Raussert, and Julia Roth, eds. *Cherishing the Past, Envisioning the Future. Entangled Practises of Heritage and Utopia in the Americas.* 2021. 176 pp.
WVT Wissenschaftlicher Verlag Trier ISBN 978-3-86821-804-6 € 23,50
UNO University of New Orleans Press ISBN 978-1-60801-206-0 $ 27.50

Vol. 26
Raussert, Wilfried. *'What's Going On': How Music Shapes the Social.* 2021. 224 pp.
WVT Wissenschaftlicher Verlag Trier ISBN 978-3-86821-811-4 € 28,50
UNO University of New Orleans Press ISBN 978-1-60801-199-5 $ 34.00

Vol. 27
Frank-Job, Barbara. *Immigration as a Process: Temporality Concepts in Blogs of Latin American Immigrants to Québec.* 2021. 138 pp.
WVT Wissenschaftlicher Verlag Trier ISBN 978-3-86821-820-6 € 20,00
UNO University of New Orleans Press ISBN 978-1-60801-215-2 $ 24.00

Vol. 28
Roth, Julia. *Can Feminism Trump Populism? Right-Wing Trends and Intersectional Contestations in the Americas.* 2021. 166 pp.
WVT Wissenschaftlicher Verlag Trier ISBN 978-3-86821-821-3 € 23,00
UNO University of New Orleans Press ISBN 978-1-60801-205-3 $ 26.00

Vol. 30
Buitrago Valencia, Clara. *Missionaries: Migrants or Expatriates? Guatemalan Pentecostal Leaders in Los Angeles.* 2021. 236 pp.

WVT Wissenschaftlicher Verlag Trier	ISBN 978-3-86821-818-3	€ 28,50
UNO University of New Orleans Press	ISBN 978-1-60801-210-7	$ 34.50

Vol. 31
Schwabe, Nicole. *De-Centering History Education: Creating Knowledge of Global Entanglements.* 2021. 92 pp.

WVT Wissenschaftlicher Verlag Trier	ISBN 978-3-86821-828-2	€ 18,00
UNO University of New Orleans Press	ISBN 978-1-60801-214-5	$ 21.00

Vol. 32
Manke, Albert. *Coping with Discrimination and Exclusion. Experiences of Free Chinese Migrants in the Americas in a Transregional and Diachronic Perspective.* 2021. 162 pp.

WVT Wissenschaftlicher Verlag Trier	ISBN 978-3-86821-829-9	€ 23,00
UNO University of New Orleans Press	ISBN 978-1-60801-207-7	$ 27.00

Vol. 33
Rohland, Eleonora. *Entangled Histories and the Environment? Socio-Environmental Transformations in the Caribbean, 1492-1800.* 2021. 92 pp.

WVT Wissenschaftlicher Verlag Trier	ISBN 978-3-86821-833-6	€ 18,00
UNO University of New Orleans Press	ISBN 978-1-60801-208-4	$ 21.00

Vol. 34
Kaltmeier, Olaf. *National Parks from North to South. An Entangled History of Conservation and Colonization in Argentina.* 2021. 210 pp.

WVT Wissenschaftlicher Verlag Trier	ISBN 978-3-86821-834-3	€ 27,50
UNO University of New Orleans Press	ISBN 978-1-60801-204-6	$ 32.50

Vol. 35
Raussert, Wilfried. *Off the Grid. Art Practices and Public Space.* 2021. 232 pp.

WVT Wissenschaftlicher Verlag Trier	ISBN 978-3-86821-835-0	€ 29,50
UNO University of New Orleans Press	ISBN 978-1-60801-213-8	$ 34.50

Vol. 36
Ravasio, Paola. *This Train Is Not Bound to Glory. A Study of Literary Trainscapes.* 2021. 114 pp.

WVT Wissenschaftlicher Verlag Trier	ISBN 978-3-86821-836-7	€ 18,00
UNO University of New Orleans Press	ISBN 978-1-60801-216-9	$ 21.00

Vol. 37

Schäfer, Heinrich Wilhelm. *Protestant 'Sects' and the Spirit of (Anti-)Imperialism. Religious Entanglements in the Americas.* 2021. 242 pp.

WVT Wissenschaftlicher Verlag Trier	ISBN 978-3-86821-855-9	€ 29,50
UNO University of New Orleans Press	ISBN 978-1-60801-209-1	$ 34.50